SCHOLASTIC

English & Maths Bumper Book Ages 8–9

2 in 1

Revision & Practice

KS2 Year 4

Master maths and English topics with ease

SCHOLASTIC

First published in the UK by Scholastic, 2016;
this edition published 2023

Scholastic Distribution Centre, Bosworth Avenue, Tournament
Fields, Warwick, CV34 6UQ

Ireland, 89E Lagan Road, Dublin Industrial Estate,
Glasnevin, Dublin, D11 HP5F

www.scholastic.co.uk

A CIP catalogue record for this book is available from
the British Library.

ISBN 978-0702-32676-9
Printed and bound by Bell and Bain Ltd, Glasgow

The book is made of materials from well-managed, FSC-certified
forests and other controlled sources.

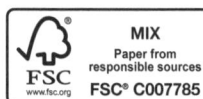

MIX
Paper from
responsible sources
FSC® C007785

FSC
www.fsc.org

Due to the nature of the web we cannot guarantee the content or
links of any site mentioned.

We strongly recommend that teachers check websites before
using them in the classroom.

Consultants

Lesley and Graham Fletcher (English)

Authors

Catherine Casey (English) and Paul Hollin (Maths)

Editorial team

Rachel Morgan, Vicki Yates, Audrey Stokes, Tracey Cowell,
Maggie Donovan, Shelley Welsh, Helen Lewis, Jenny Wilcox,
Mark Walker, Mary Nathan, Christine Vaughan, Kate Pedlar,
Janette Ratcliffe and Julia Roberts

Design team

Dipa Mistry, Andrea Lewis, Nicolle Thomas, Neil Salt and Oxford
Designers and Illustrators

Illustration

Simon Walmesley and Matt Ward @ Beehive Illustration

Contents

Maths Made Simple

How to use

This book has been written to help children reinforce the English and maths skills they have learned in school. Each subject is divided into sections covering a range of topics from the National Curriculum. Use the book little and often to practise skills and increase confidence. You can choose to work through the English and maths sections in order or focus on specific topics.

At the back of the book is a **Progress tracker** to enable you to record what has been practised and achieved.

English

Grammatical words

Verb tenses 2

↻ Recap 3

1

What are verb tenses?

Verbs are doing or being words. They describe what is happening. Verbs come in different **tenses** which tell us when something happened.

- **Present tense** – for events that are happening now.
- **Past tense** – for events that have happened.
- **Present progressive** – for events that are still happening.
- **Past progressive** – for events that were happening over a period of time.
- **Present perfect** – for events that started in the past but they have a known end.

KEY WORDS 7

verbs
tense
past tense
present tense
past progressive
present progressive
present perfect

📝 Revise 4

Look at the verb tenses and forms highlighted in these examples.

I run to catch the bus.	← present tense – the action is happening
I ran to catch the bus.	← past tense – the action has happened
I am running to catch the bus.	← present progressive – use a helper verb (am/is) and the verb+ing
I was running to catch the bus.	← past progressive – use a helper verb (was/were) and the verb+ing
I have run to catch the bus.	← present perfect – use 'have' + verb

💡 Tip 5

Watch out for irregular verbs such as run, swim, draw, cut, know, take and write. These verbs don't add the suffix ed for the past tense.

✔ Skills check 6

1. Choose and fill in the correct word to complete each sentence in the past tense.

   a. We _____ at the outdoor pool in town. (**swam/swimmed**)

   b. He _____ a letter of complaint. (**writted/wrote**)

   c. The frog _____ out of the pond. (**jumped/jamp**)

2. Draw lines to match each sentence to the correct verb type.

Sentence	Verb type
I have drawn a picture.	Past tense
I am drawing a picture.	Present tense
I draw a picture.	Past progressive
I was drawing a picture.	Present progressive
I drew a picture.	Present perfect

3. Fill in the gaps in the sentence below, using the past progressive form of the verbs in the boxes.

   to play   to cook

   I _____ with my toys while Dad _____ the dinner.

4. Rewrite this sentence in the present perfect.

   The princess **rescued** the prince from the tower.

   _____

14   15

1. Chapter title
2. Topic title
3. Each page starts a **recap** with a 'What is…' question which gives children a clear definition for the terminology used.
4. In the **revise** section there are clear explanations and examples, using clear illustrations and diagrams, where relevant.
5. **Tips** provide short and simple advice to aid understanding.
6. The **skills check** sections enable children to practise what they have learned with answers at the back of the book.
7. **Key words** that children need to know are displayed. Definitions for these words can be found in the **Glossary**.

Maths

The Maths section has many of the same features of the English section and also some additional ones. Keep some blank or squared paper handy for notes and calculations!

Money **2**

↻ Recap **3**

These are the coins we use in England and Wales.
We also use £5, £10, £20 and £50 notes.

1

📖 Revise **4**

Money shows us the cost of things. We use pounds and pence.
£1 = 100 pence

We show pence using two decimal places.
7 pounds and 25 pence = £7.25
That's *seven pounds twenty-five*.

Unlike other decimals, if the last digit is a zero, we still write it in.
16 pounds and 50 pence = £16.50
That's *sixteen pounds fifty*.

Look at this amount: £0.59 is zero pounds and fifty-nine pence, or 59p

To convert pounds to pence, multiply by 100:
£6.50 = 6.50 × 100 = 650p

To convert pence to pounds divide by 100:
3265p = 3265 ÷ 100 = £32.65

DID YOU KNOW?
Before the year 1971 we used pounds, shillings and pence. A shilling was worth 12 old pennies and there were 240 old pennies in a pound!

1p is one hundredth of one pound.

Notice that if you use the £ sign and decimals, you don't add a p at the end.

💡 Tips **5**

Operation	Example
Addition	£3.50 + £2.15 = £5.65
Subtraction	£5.00 − £1.25 = £3.75
Multiplication	£2.10 × 3 = £6.30
Division	£7.00 ÷ 2 = £3.50
Fractions	$\frac{1}{2}$ of £25.00 = £12.50

You can use all your number skills to solve money problems.

- We can use written methods with money just like any other numbers.
- Remember to be careful with the decimal point.

💬 Talk maths **6**

Find an old shopping receipt or a price list from a catalogue or website, and work with a partner to compare costs.

Next, challenge each other by asking for items on the list and paying for them.

If you feel confident, put the money away and solve problems just using the maths.

I'd like a tin of beans for 32p. Here is £1.

Here is your change: 68p.

✔ Check **7**

1. Convert these pence to pounds.

Pence	500p	150p	3300p	59p	1000p
Pounds					

2. Convert these pounds to pence.

Pounds	£1	£4.25	£0.62	£20	£12.06
Pence					

3. Complete these calculations.

   a. £2.50 + £3.30 = _____   b. £4.90 + £3.20 = _____
   c. £10.00 − £6.50 = _____   d. £20.00 − £12.99 = _____
   e. 8 × 50p = £ _____   f. £2.50 × 4 = £ _____
   g. £20 ÷ 4 = £ _____   h. £15 ÷ 3 = £ _____

⚠ Problems **8**

Brain-teaser Ice creams cost £1.25 each. Alfie's mum buys five.

a. What is the total cost? _____

b. How much change will she get from a £10 note? _____

Brain-buster Ice creams cost £1.25 each and ice lollies cost £1.50 each.

a. How much would three ice creams and seven lollies cost altogether? _____

b. How much change would there be from a £20 note? _____

112   113

---

1. Chapter title
2. Topic title
3. Each page starts a **recap** of basic facts of the mathematical area in focus.
4. In the **revise** section there are clear explanations and examples, using illustrations and diagrams, where relevant.
5. **Tips** provide short and simple advice to aid understanding.
6. **Talk maths** are focused activities that encourage verbal practice.
7. **Check** a focused range of questions, with answers at the end of the book.
8. **Problems** word problems requiring mathematics to be used in context.

Tips for using this book at home

Using this book, alongside the maths and English being done at school, can boost children's mastery of the concepts. Be sure not to get ahead of schoolwork or to confuse your child.

Keep sessions to an absolute maximum of 30 minutes. Even if children want to keep going, short amounts of focused study on a regular basis will help to sustain learning and enthusiasm in the long run.

Word lists These are the words you need to learn to spell.

Years 3–4

accident	difficult	interest	potatoes
accidentally	disappear	island	pressure
actual	early	knowledge	probably
actually	earth	learn	promise
address	eight/eighth	length	purpose
answer	enough	library	quarter
appear	exercise	material	question
arrive	experience	medicine	recent
believe	experiment	mention	regular
bicycle	extreme	minute	reign
breath	famous	natural	remember
breathe	favourite	naughty	sentence
build	February	notice	separate
busy/business	forward/forwards	occasion	special
calendar	fruit	occasionally	straight
caught	grammar	often	strange
centre	group	opposite	strength
century	guard	ordinary	suppose
certain	guide	particular	surprise
circle	heard	peculiar	therefore
complete	heart	perhaps	though/although
consider	height	popular	thought
continue	history	position	through
decide	imagine	possess	various
describe	increase	possession	weight
different	important	possible	woman/women

Multiplication table

x	1	2	3	4	5	6	7	8	9	10	11	12
1	1	2	3	4	5	6	7	8	9	10	11	12
2	2	4	6	8	10	12	14	16	18	20	22	24
3	3	6	9	12	15	18	21	24	27	30	33	36
4	4	8	12	16	20	24	28	32	36	40	44	48
5	5	10	15	20	25	30	35	40	45	50	55	60
6	6	12	18	24	30	36	42	48	54	60	66	72
7	7	14	21	28	35	42	49	56	63	70	77	84
8	8	16	24	32	40	48	56	64	72	80	88	96
9	9	18	27	36	45	54	63	72	81	90	99	108
10	10	20	30	40	50	60	70	80	90	100	110	120
11	11	22	33	44	55	66	77	88	99	110	121	132
12	12	24	36	48	60	72	84	96	108	120	132	144

English Made Simple
Ages 8–9

Proper nouns and common nouns

What are proper nouns and common nouns?

↺ Recap

Nouns are the names of people, places and things. There are different types of noun.

A **proper noun** names something specific and a **common noun** names something in general.

📋 Revise

In this table, the nouns have been sorted into common nouns and proper nouns.

Common noun	Proper noun
month	August
planet	Saturn
building	Eiffel Tower

Look at the nouns highlighted in this example.

Jack travelled to **France** on an **aeroplane**.

proper noun (name of specific person)   proper noun (name of specific place)   common noun (name of thing)

Tip 💡

Remember, proper nouns always start with a capital letter.

✔ Skills Check

Can you find the nouns in these sentences?

1. **Underline the common nouns and circle the proper nouns in these sentences.**

   a. The astronauts prepared for their journey to Mars.

   b. Ms Green gave the class their homework.

   c. The doctor used a stethoscope to listen to Amelia's heart.

   d. The tourists visited Buckingham Palace in London.

   e. "My birthday is in June," said Hannah excitedly.

KEY WORDS

nouns
proper nouns
common nouns

Adjectives

What is an adjective?

↺ Recap

Adjectives are often called describing words. They describe features of nouns such as colour, age, shape or size.

📋 Revise

Look at the adjectives highlighted in this example. They give us more detail about the lighthouse steps.

The **ancient** lighthouse steps were **creaky** and **rotten**.

What effect do different adjectives have on the sentence?

Tip 💡

Adjectives make a sentence more interesting by providing the reader with more detail.

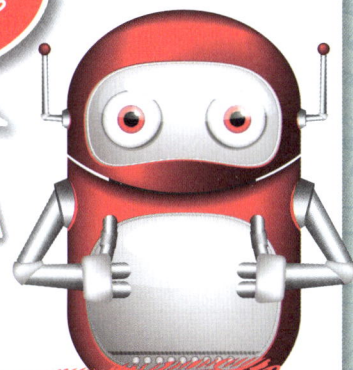

KEY WORDS

adjectives

✔ Skills Check

1. Improve these sentences by using more exciting adjectives than the ones in bold.

   a. The (**big**) _____ fish swam around the pond.

   b. The captain climbed aboard the (**old**) _____ boat.

   c. The sandwiches we ate for lunch were (**nice**) _____.

2. Rewrite these sentences, adding adjectives to make them more interesting.

   a. The owl sat on a branch.

   _____

   b. The cyclist rode down the lane.

   _____

   c. The teacher entered the hall.

   _____

Adjectives with prefixes

How do prefixes change adjectives?

↺ Recap

A **prefix** is a set of letters added to the beginning of a word. When prefixes are added to adjectives, they can change their meaning.

▤ Revise

Let's look at the prefixes in these examples and see how they change the meaning of the adjective they are added to.

Prefix	Adjective	New word
un	happy	unhappy
in	active	inactive
im	polite	impolite
ir	regular	irregular

The prefixes un, in, im and ir give the adjective the opposite meaning.

✔ Skills Check

1. Add the prefix to the adjective to create a new word. Write the new word in the box.

Prefix	Adjective	New word
un	helpful	
in	complete	
ir	responsible	

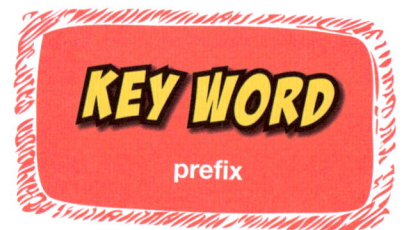

KEY WORD

prefix

2. What effect does the prefix 'un' have on these words? Explain your answer.

un + interested = **un**interested   un + safe = **un**safe

_____

_____

_____

_____

Tip 💡

Most prefixes can be added without changing the spelling of the adjective.
For example: **un** + comfortable = **un**comfortable

Noun phrases

What is a noun phrase?

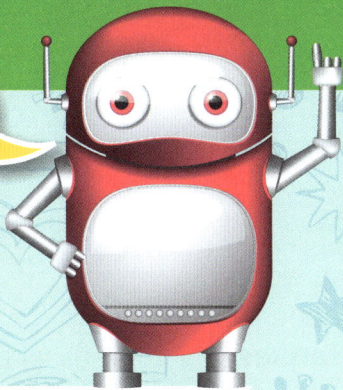

↺ Recap

A **noun phrase** has a noun as its main word and has one or more adjectives and/or a preposition (see page 26).

📝 Revise

Let's look at the expanded noun phrases highlighted in these examples.

> The **amazing new robot with three arms** cleaned the floor!
>
> adjectives   noun   preposition   noun

> I bought **the last computer-tablet in the shop**.
>
> adjective   noun   preposition   noun

> The swimmer dived into **the cold outdoor pool behind the trees**.
>
> adjectives   noun   preposition   noun

You know about adjectives, but can you see how to use prepositions to give more information?

✔ Skills Check

1. Underline the noun phrases in these sentences.

   a. The lonely, frightened evacuee with a suitcase stood on the platform.

   b. The robin stood on the broken, empty bird-bath by the path.

   c. The children played happily in the soft, yellow sand near the dunes.

   d. Amber borrowed the only English dictionary in the library.

   e. Omar took the last apple muffin on the tray.

KEY WORD

noun phrases

Verb tenses

↺ Recap

What are verb tenses?

Verbs are doing or being words.
They describe what is happening.
Verbs come in different **tenses** which tell us when something happened.

- **Present tense** – for events that are happening now.
- **Past tense** – for events that have happened.
- **Present progressive** – for events that are still happening.
- **Past progressive** – for events that were happening over a period of time.
- **Present perfect** – for events that started in the past but they have a known end.

KEY WORDS

verbs
tense
past tense
present tense
past progressive
present progressive
present perfect

📝 Revise

Look at the verb tenses and forms highlighted in these examples.

I **run** to catch the bus. ← present tense – the action is happening

I **ran** to catch the bus. ← past tense – the action has happened

I **am running** to catch the bus. ← present progressive – use a helper verb (am/is) and the verb+ing

I **was running** to catch the bus. ← past progressive – use a helper verb (was/were) and the verb+ing

I **have run** to catch the bus. ← present perfect – use 'have' + verb

💡 Tip

Watch out for irregular verbs such as **run**, **swim**, **draw**, **cut**, **know**, **take** and **write**.
These verbs don't add the suffix **ed** for the past tense.

✔ Skills Check

1. Choose and fill in the correct word to complete each sentence in the past tense.

   a. We _____ at the outdoor pool in town. (**swam/swimmed**)

   b. He _____ a letter of complaint. (**writted/wrote**)

   c. The frog _____ out of the pond. (**jumped/jamp**)

2. Draw lines to match each sentence to the correct verb type.

Sentence	Verb type
I have drawn a picture.	Past tense
I am drawing a picture.	Present tense
I draw a picture	Past progressive
I was drawing a picture.	Present progressive
I drew a picture.	Present perfect

3. Fill in the gaps in the sentence below, using the past progressive form of the verbs in the boxes.

   to play   to cook

   I _____ with my toys while Dad _____ the dinner.

4. Rewrite this sentence in the present perfect.

   The princess **rescued** the prince from the tower.

   _____

Adverbs

What is an adverb?

↻ Recap

Adverbs describe verbs. They tell you more about the event or action, describing how, when, where or why something happens.

🗎 Revise

Here are some different types of adverb.

How something happens	When or how often something happens	Where something happens	Why something happens
unexpectedly	today	here	therefore
powerfully	now	there	otherwise
accidentally	soon	everywhere	consequently
gracefully	never	indoors	hence
mysteriously	often	upstairs	
gradually	regularly	abroad	

Look at the **adverbs** highlighted in these examples and how they describe the **verb**.

Mysteriously, my socks have **disappeared**. ← describes how the socks have disappeared

I **go** swimming **regularly**. ← describes how often I go swimming

They **looked everywhere** for the ball. ← describes where they looked

💡 Tip

Adverbs can be at the beginning, middle or end of a sentence.

KEY WORD

adverbs

✔ Skills Check

1. **Circle the adverb in each sentence below.**

   a. Sadly, James picked up the broken doll.

   b. We sometimes play tennis.

   c. The children are playing downstairs.

   d. I went shopping yesterday.

   e. Suddenly, the cat leapt off the wall.

2. **Complete these sentences with a suitable adverb.**

   a. _____, the children splashed in the paddling pool together.

   b. The dog barked _____ .

   c. _____, I am going to the cinema.

   d. The cat curled up _____ after a night hunting mice.

   e. _____, the dentist removed the tooth.

3. **Replace the adverbs that are underlined in the sentences below. Rewrite each sentence.**

   What effect does changing the adverb have?

   a. Flora broke the precious glass <u>accidentally</u>.

   _____

   b. Nikhil was playing <u>indoors</u>, with the ball.

   _____

   c. <u>Often</u>, I get the bus to school.

   _____

Adverbials

↺ Recap

An **adverbial** is a group of words or a phrase that behaves like an adverb. It tells you more about the event or action such as how, when, where or why.

📄 Revise

These two examples show how adverbs and adverbials do the same 'job' in a sentence.

I **go** swimming **regularly**.

↑

adverb describes how often I go swimming

I **go** swimming **every Monday and Thursday**.

↑

adverbial describes how often I go swimming

Look at the different types of adverbial highlighted in these examples.

Isabel **ate** her dinner **as slowly as she could**.

↑

adverbial: tells you how Isabel ate

The cricket team **had** oranges **at half time**.

↑

adverbial: tells you when the cricket team had oranges

We **had** a barbeque **on the patio**.

↑

adverbial: tells you where we had a barbeque

The mouse **fled because of the cat**.

↑

adverbial: tells you why the mouse fled

✔ Skills Check

1. **Underline the adverbial in each sentence below.**

   a. I went to the park last Thursday afternoon.

   b. We waited for our drinks in the sunshine.

   c. The little girl ran to the finishing line as fast as she could.

   d. The children were playing football all morning.

   e. The enormous dog barked in the garden.

KEY WORD

adverbials

18

Fronted adverbials

↺ Recap

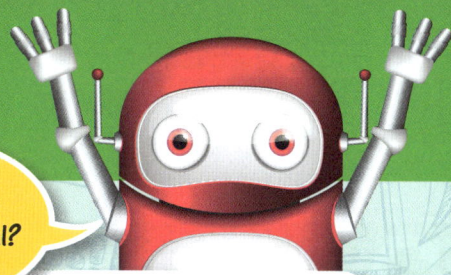

What is a fronted adverbial?

Fronted adverbials are adverbials that are at the beginning of a sentence.

📄 Revise

Let's see how the adverbials highlighted in these examples can move to the front of the sentence. Fronted adverbials are usually followed by a comma.

The baby slept **in his cot**.   **In his cot**, the baby slept.

⬆   ⬆   ⬆

adverbial   fronted adverbial   comma

both describe where the baby slept

We went bowling **the day before yesterday**.

⬆

adverbial

The day before yesterday, we went bowling.

⬆   ⬆

fronted adverbial   comma

both describe when they went bowling

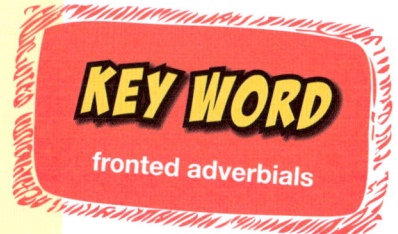

KEY WORD

fronted adverbials

✔ Skills check

1. Rewrite the sentences below so that they begin with the adverbial in bold. Use only the same words.

a. Ms Wilkinson played the piano **in assembly**.

_____

b. The bell rang **suddenly**.

_____

c. We went inside **at the end of break time**.

_____

d. The woman walked her dog **along the beach**.

_____

Clauses

↻ Recap

What is a clause?

A **clause** is a group of words that contains a verb and tell you who or what is doing the verb. Clauses can sometimes be complete sentences.

- A **main clause** makes sense on its own, tells you who or what does the verb and has a verb.
- A **subordinate clause** needs the rest of the sentence (a main clause) to make sense.
  A subordinate clause includes a conjunction to link it to the main clause.

🗒 Revise

Look at the main and subordinate clauses highlighted in these examples. A main clause can be a sentence on its own.

She is going out.
↑
main clause

She is going out after she has eaten her dinner.
↑   ↑   ↑
main clause –   conjunction   subordinate clause
makes sense   – does not make
by itself   sense by itself

When I picked it up, the hamster tickled my hands.
↑   ↑   ↑
conjunction   subordinate   main clause –
            clause – does   makes sense
            not make sense   by itself
            by itself

KEY WORDS

clause
main clause
subordinate clause

💡 Tips

Subordinate clauses can be at the beginning or end of a sentence.

Can you tell if these clauses are main clauses or subordinate clauses?

✔ Skills Check

1. Put a tick in each row to show whether the main clause or the subordinate clause is in bold.

	Main clause	Subordinate clause
I washed my hands **after I went to the toilet**.		
When I lost my favourite teddy, I was upset.		
I jumped when the door slammed loudly.		
Before I went on stage, **I was feeling nervous**.		
I shut the window when it rained.		
The lights went out **because the power was cut off**.		

2. Underline the main clause in each sentence below.

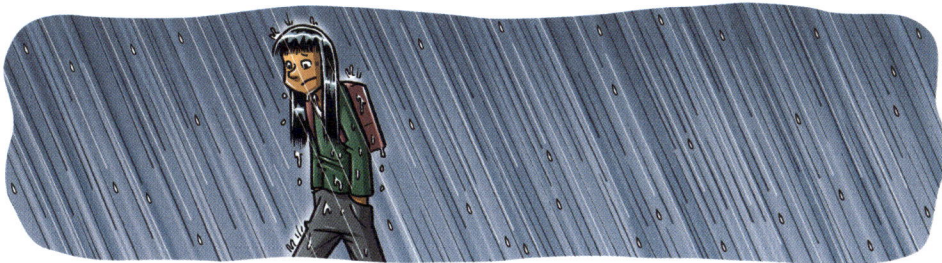

   a. The girl walked to school although it was raining.

   b. The bus was late because it broke down.

   c. As it was snowing, the football match was cancelled.

3. Underline the subordinate clause in each sentence below.

   a. He cleaned out the guinea pigs after feeding the rabbit.

   b. When we were younger, we went ice-skating with our grandma.

   c. If I go on Saturday, I will see the animals at the zoo.

Conjunctions

↻ Recap

What is a conjunction?

Conjunctions join words or clauses together.

- **Co-ordinating conjunctions** link two main clauses together.

They include: **and** **but** **or** **so**

- **Subordinating conjunctions** link a subordinate clause to a main clause. They include: **when** **if** **after** **because** **before**

📑 Revise

Look at the co-ordinating and subordinating conjunctions highlighted in these examples.

I enjoy swimming **but** I don't like diving.

↑ main clause   ↑ co-ordinating conjunction   ↑ main clause

She went to the park **because** she wanted to go on the swings.

↑ main clause   ↑ subordinating conjunction   ↑ subordinate clause

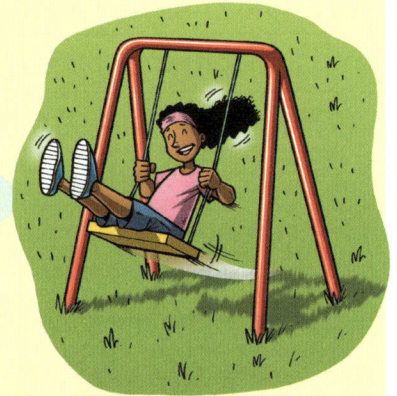

When you knocked on the door, I opened it.

↑ subordinating conjunction   ↑ subordinate clause   ↑ main clause

KEY WORDS

conjunctions
co-ordinating conjunctions
subordinating conjunctions

✔ Skills Check

1. Circle the co-ordinating conjunctions and underline the subordinating conjunctions in the sentences below.

a. She put sun cream on before she went outside.

b. I have two brothers so I know lots about football.

c. You can have raisins or you can have grapes.

d. If you cook dinner, I'll do the washing up.

Determiners

What is a determiner?

↻ Recap

Determiners go before a noun (or noun phrase) and show which noun you are talking about. They can tell you whether the noun is known or unknown, who it belongs to or how many there are. Determiners include: **the   a   an   your   my   some   every**

KEY WORD
determiners

目 Revise

Determiners tell you…	Examples	Sentences
whether the noun is known or not	**a** or **an** (unknown) **the** (known)	I saw **an** umbrella lying in the corridor. **The** umbrella was blue.
who the noun belongs to	**your**, **my**, **their**, **her**	Can I borrow **your** coat? **My** coat is still wet.
how many of the noun	**some**, **every**, **one**, **two**	**Some** children forgot their homework. **Two** cats were sitting on the wall.

Here are some examples of determiners in use.

"Can you get me **a** new library **book**?" asked Thomas.

determiner: **a** (an unknown noun – could be any book)

determiner

"They've got **some** great books on wizards!" enthused Arthur. "**My** favourite one is **the book** on Italian cooking."

determiner

determiner: **the** (a known noun – Italian cooking book)

💡 Tip

When you're using **a/an**:
- use **a** if the noun (or noun phrase) begins with a consonant
- use **an** if the noun (or noun phrase) begins with a vowel

✔ Skills Check

1. Circle the determiner in each sentence below.

   a. I need to see a dentist.   **b.** Please put on your socks and shoes.

2. Write the correct determiner to complete these sentences – a, an or some.

   a. The boy ate _____ orange for his lunch.   **b.** Please can I have _____ peas?

Pronouns

What is a pronoun?

Pronouns are words used instead of nouns or noun phrases. It means that the noun does not need to be repeated.

📄 Revise

Pronouns replace nouns to avoid repetition. They include:

I   she   he   it   you   we   they

Noun	Pronoun
Amira was late for school. **Amira** ran down the road.	Amira was late for school. **She** ran down the road.
The children were in the garden. **The children** were playing football.	The children were in the garden. **They** were playing football.

Possessive pronouns are used to show who or what something belongs to. They include:

mine   hers   his   its   yours   ours   theirs

Noun	Pronoun
The teacher put on **the teacher's** jacket.	The teacher put on **his** jacket.
That book is **Mrs Newley's**.	That book is **hers**.
The cat has drunk all **the cat's** milk.	The cat has drunk all **its** milk.

💡 Tip

Be careful when you use pronouns – make sure it's clear which noun they refer to.

✔ Skills Check

1. Circle all the pronouns in the sentences below.

   a. Ruby and Edward went to the park. They played football.

   b. I played the guitar loudly.

   c. Connor finished his homework quickly.

   d. Sarah was doing well at swimming. She didn't need her armbands any more.

   e. Darcy had PE but she couldn't find her kit.

2. **Complete the sentences by filling in the missing pronouns.**

   his   its   she   they   he

   a. Meesha was going to a party but _____ didn't like her dress.

   b. Felix was hungry because _____ had forgotten his packed lunch.

   c. The alien landed _____ spacecraft on the planet Earth.

   d. David and Rosie put on sun cream but _____ didn't wear their sun hats.

   e. Vishal searched everywhere but _____ couldn't

   find _____ school shoes.

Prepositions

↻ Recap

What is a preposition?

Prepositions link nouns, pronouns or noun phrases to other words in the sentence.

Prepositions usually tell you about place, direction or time. They include:

before   after   during   in   because of

under   on   around   beside

KEY WORD
prepositions

Revise

Look at the prepositions highlighted in these examples.

I wanted the shiny green scooter **in** the shop window.

Can I sleep **on** the top bunk bed?

I ran **around** the corner.

✔ Skills Check

1. Circle the prepositions in these sentences.

a. We swam after lunch.

b. I needed the toilet during assembly.

c. The astronaut landed beside a crater.

d. The lorry went under the bridge.

2. Choose and fill in the best preposition to complete each sentence below.

a. School was closed _____ the snow. (**because of/on**)

b. I had to eat all my carrots _____ pudding. (**in/before**)

c. The snake was curled _____ a branch of the tree. (**around/after**)

Capital letters

↺ Recap

When do you use a capital letter?

Use a capital letter:
- to mark the beginning of a sentence
- for proper nouns, including people's names, days of the week and names of places (such as cities, restaurants and shops).

目 Revise

Look at the capital letters highlighted in this example.

(T)he (F)rench chef rustled up a delicious lunch for (A)ndrew.

to show the beginning of a sentence

to show proper nouns (nationality and a boy's name)

Tip

Remember to use a capital letter for the pronoun 'I'.

✔ Skills Check

1. Rewrite these sentences, putting capital letters in the correct places.

   a. usually in october, the leaves fall off the trees.

   _____

   b. lily and meg visited edinburgh castle, on their school trip.

   _____

   c. the new pilot, who was called tom, often flew to germany.

   _____

2. **Elijah says, "Capital letters are only used at the beginning of a sentence."**

   Is he right? Circle the correct answer.   Yes   No

   Explain your answer.

   _____

Full stops, question marks and exclamation marks

When do you use a full stop, question mark or exclamation mark?

↺ Recap

Full stops mark the end of statements or commands.

Question marks show the end of questions.

Exclamation marks can be used to show strong feelings such as surprise or panic. They can also be used to indicate that someone is shouting or that something is very loud.

. full stop
? question mark
! exclamation mark

Revise

"What fantastic writing today(!)" exclaimed Ms Black.

exclamation mark: to show the teacher's strong feelings about how well Lucy has done

"Really(?)" asked Lucy, wide-eyed.

question mark: to show Lucy is asking a question

"Yes, all the letters are joined clearly," replied Mrs Black, smiling(.)

full stop: to show the end of a sentence

KEY WORDS

full stops
question marks
exclamation marks

💡 Tips

- Full stops, question marks and exclamation marks help the reader to use the correct expression and emphasis when reading.
- Exclamation marks are often used when writing direct speech.

✔ Skills Check

1. a. Which sentence is punctuated most appropriately? Tick one.

When it rains, our garden is full of puddles. ☐

When it rains, our garden is full of puddles! ☐

When it rains, our garden is full of puddles? ☐

b. Explain your answer.

_____

_____

2. Complete each sentence with the most appropriate punctuation mark.

a. When can I go outside to play ◯

b. When it rains hard, we have break time inside ◯

c. Pick that up ◯

d. What an amazing feeling ◯

3. In the passage below, some of the punctuation is missing. Add the correct punctuation in each of the spaces.

. ! ?

It was the school holidays ◯ Jessica and Nathan were out walking in the woods when they came to a broken bridge.
"How annoying ◯" cried Nathan.
"How are we going to get across the stream now ◯" thought Jessica.
They both looked at the fast-flowing stream and the slippy rocks underneath the broken bridge ◯ It was no good, they would have to walk downstream until they found a safe place to cross.

Apostrophes for contraction

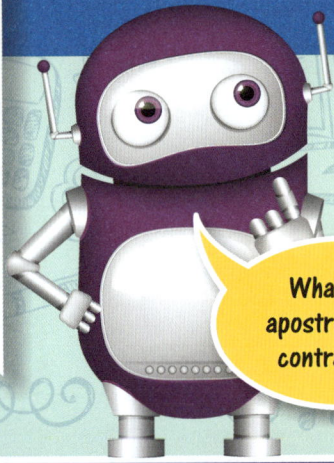

↻ Recap

What is an apostrophe for contraction?

An **apostrophe** is a punctuation mark: '

Apostrophes for **contraction** are used to show the place of a missing letter or letters when two words are joined (for example, **that's** for **that is**).

Revise

Look at the examples. The apostrophe goes where the missing letters are.

Words in full	Contraction	Sentence
it is	it's	**It's** a long way to Grandad's house.
they are	they're	**They're** going to win the race.
could not	couldn't	She **couldn't** reach the top shelf.
he will	he'll	**He'll** have to look in the lost property box.
I have	I've	**I've** finished my work.
what is	what's	**What's** the time?
you are	you're	**You're** swimming in the main pool tonight.

💡 Tip

When you see a contraction, think about what it *means*, so you can work out what the missing letters are.

✔ Skills Check

KEY WORDS

apostrophes
contraction

1. Rewrite these words as contractions using an apostrophe.

does not   He is

a. He _____ want to go to school today. _____ unwell.

cannot   I will

b. I _____ remember my password. _____ have to reset it.

2. Rewrite these contractions in full.

Words in full	Contraction	Words in full	Contraction
	you'd		should've
	aren't		he'll

Apostrophes for possession

↻ Recap

What is an apostrophe for possession?

Apostrophes for **possession** show who or what something belongs to.

📄 Revise

If something belongs to one person then you add **'s**.

> I borrowed **Tracey's umbrella** to go out in the rain.
> ↑
> shows that the umbrella belonged to Tracey

Who do the items belong to in these sentences?

If it is an irregular plural that doesn't end with **s**, then add **'s**.

> The cloakroom was a mess because the **children's coats** were all over the floor.
> ↑
> shows that the coats belonged to the children

If something belongs to more than one person then you add an apostrophe after the s.

> It was the **girls' turn** to go on the trampoline.
> ↑
> shows that the turn belonged to the girls (plural – more than one girl)

✔ Skills Check

1. Circle the word in each sentence that should contain an apostrophe. Then rewrite it in the box to show that there is more than one owner in each case.

   a. After PE, the childrens school shoes were all muddled up. [   ]

   b. The bridesmaids dresses had not arrived. [   ]

   c. The babies mouths were wide open in surprise. [   ]

   d. All the boys costumes were ready. [   ]

KEY WORDS

possession

Inverted commas

What are inverted commas?

Inverted commas are punctuation marks that show **direct speech**: " "

Sometimes they are also called speech marks.

📄 Revise

Inverted commas go at the beginning and end of direct speech. They enclose the spoken words and the punctuation that goes with the speech. Who is talking and how comes outside of the inverted commas.

"Tomorrow I'm going to my friend's house," said Sophia.

inverted commas

direct speech

final comma: inside the inverted commas

who is talking and how

Tomorrow I'm going to my friend's house.

"Which way is the post office, please?" asked the old man.

inverted commas

question mark: inside the inverted commas

Which way is the post office, please?

"I'm over here!" shouted Jakub.

inverted commas

exclamation mark: inside the inverted commas

I'm over here!

The teacher told the children, **"Line up now."**

who is talking and how first, followed by a comma: before the inverted commas

comma

inverted commas

Line up now.

Tips

- Use a comma after who is talking, when it comes before the direct speech.
- When you're writing down a conversation, start a new line each time the speaker changes.

✔ Skills Check

1. **Which sentence uses inverted commas correctly? Tick one.**

"Look at all that rain"! exclaimed Grandad. "I think we will have to go in the car today". ☐

"Look at all that rain! exclaimed Grandad. I think we will have to go in the car today." ☐

"Look at all that rain!" exclaimed Grandad. "I think we will have to go in the car today." ☐

"Look at all that rain!" exclaimed Grandad. "I think we will have to go in the car today". ☐

2. **In the passage below, insert the missing inverted commas.**

Charlie was standing at the end of the dinner queue.

I am so hungry! he moaned.

Me too. Why are we always last? said his friend Sam.

I just hope there is some chocolate cake left, replied Jing, who was just in front of Charlie.

Then the lunchtime assistant told them, You don't need to worry. There's plenty of cake for everyone.

Commas in lists

↺ Recap

When and how do you use commas in a list?

Commas can be used to separate items in a list instead of repeating the word 'and'.

🖹 Revise

You put a comma between each item in the list except the last item, where you use 'and'. An item in a list may be a single word or several words.

Look at these examples to see how it's done.

> At the school fete there was a **bouncy castle, cake stalls, live music and some games.**

> For lunch I had **delicious sandwiches, a bunch of green grapes and the very last biscuit.**

💡 Tip

You don't need to use a comma if there are only two words in the list.
For example: The boy had a football and a lunchbox.

✔ Skills Check

1. **Which sentence uses commas in a list correctly? Tick one.**

   He used, flour, sugar, butter and eggs to make a delicious cake. ☐

   He used flour, sugar, butter and eggs to make a delicious cake. ☐

   He used flour, sugar, butter, eggs to make a delicious cake. ☐

   He used flour, sugar, butter, and eggs to make a delicious cake. ☐

2. **Insert the missing commas to complete these sentences.**

   a. For breakfast, I had pancakes yoghurt fruit and honey.

   b. On sports day, she competed in the egg and spoon race the skipping race and the obstacle race.

   c. In the film about nocturnal animals there were owls bats and foxes.

KEY WORD

commas

Commas after fronted adverbials

↺ Recap

What are commas after fronted adverbials?

A **fronted adverbial** is an adverbial placed at the beginning of a sentence. It is usually followed by a comma.

An adverbial tells you how, when, where or why.

🖹 Revise

Look at the fronted adverbials and commas highlighted in these examples.

In assembly, I played the recorder.

↑ fronted adverbial   ↖ comma

After break time, he had a French lesson.

↑ fronted adverbial   ↖ comma

Tomorrow morning, we are going on holiday.

↑ fronted adverbial   ↖ comma

✔ Skills check

1. Add a comma in the correct place to the sentences below.

   a. Before school I had a swimming lesson.

   b. Last year my teacher was Mr Davies.

   c. At the weekend her aunt came to visit.

KEY WORD

fronted adverbials

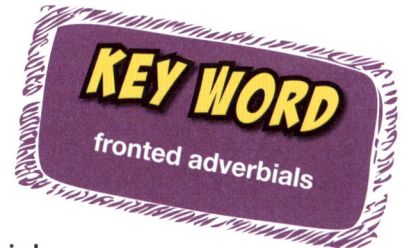

2. Rewrite each sentence below so that it begins with the adverbial. Use only the same words. Don't forget the comma.

   a. It snowed and snowed in January.

   _____

   b. I went to the museum yesterday afternoon.

   _____

   c. We had to sit and wait at the airport for Granny's plane.

   _____

Paragraphs

What is a paragraph?

↺ Recap

A paragraph is a group of sentences about the same topic. Paragraphs make text easier to read by breaking it into smaller sections. They are used to organise ideas around a common theme.

📄 Revise

A new paragraph should start when the topic changes. The first sentence of each paragraph should indicate what the paragraph is about. For example, look at the paragraphs in the passage below.

💡 Tip

You can use pronouns within and across paragraphs to link information together without repeating the noun. Make sure it's clear which noun you're referring to.

Pond dipping

On Wednesday, **Class 6** went pond dipping. We had to walk up to the pond which was along a main road. Mr Hall put us into pairs and gave us reflective jackets to wear to keep us safe. I was excited about the creatures we might find. We had been learning about habitats in Science lessons.

first paragraph: about the journey

We needed lots of special equipment for pond dipping. We used strong dip nets to catch the wildlife, white trays to put our catches in so we could see any creatures easily, magnifying glasses to look at the creatures carefully and charts to identify the insects.

the pronoun **we** refers to **Class 6** in the previous paragraph

second paragraph: about the equipment

I caught a tadpole which had two legs. There was also frog spawn in the pond. The frog spawn looked like little balls of white jelly with tiny black dots inside. Mr Hall told us that tadpoles develop tiny teeth! My tadpole had two legs so was probably about 9 weeks old.

first sentence tells you what the paragraph is about

third paragraph: about the tadpole caught during pond dipping

✔ Skills Check

1. **Read this passage and then answer the questions below.**

Then I saw a huge dragonfly high up above the water. It was much bigger than I expected. Dragonflies can be beautiful colours. The one we spotted by the pond was blue and dark green. It had enormous eyes that covered the whole of its head. After that, I stuck the dip net right to the very bottom of the pond. I got my coat a bit wet but I caught some flatworms. Flatworms live on the bottom of the pond, which I already knew because we learned about it in class. When we had looked at the creatures with our magnifying glasses, we did sketches and labelled them. Mr Hall made us put most of the creatures back in the pond, but we were allowed to take the tadpoles back to class.

a. Draw a line like this **/** in the text to show where any new paragraphs should go.

b. Explain why you think the paragraph breaks should go there.

c. Describe what each of your paragraphs is about.

Headings and subheadings

↻ Recap

What is a heading and subheading?

Headings are titles for a whole piece of text. Subheadings are titles for sections or paragraphs of text. They tell you what the section is about. They help to organise information on the page and make it easier for the reader to find information.

📄 Revise

Look at the use of headings in this example passage.

The seaside in Victorian times

Introduction
During Victorian times, the British seaside became a popular holiday destination. Families would take day trips to the beach. It was much less common to holiday abroad.

Punch and Judy
Punch and Judy was a famous puppet show performed for children on the beach.

Buckets and spades
Buckets and spades were used to build sandcastles just as they are today. However, the buckets and spades were made from metal and wood rather than the plastic used now.

The sea
The outfits people wore to swim in the sea were very different to the swimming costumes used today. Ladies wore swimwear that looked more like a dress and men usually wore an all-in-one garment that looked like long shorts and a T-shirt.

💡 Tip

Headings are usually in bold or bigger writing so that you can identify them easily. Sometimes they are underlined.

✔ Skills Check

1. **Read the passage above.**

   a. What are the four subheadings in the passage?

   _____ _____ _____ _____

   b. How do you know they are subheadings?

   _____

   c. Read the last paragraph. Explain why 'The sea' is *not* a good heading.

   _____

Word families

↻ Recap

What is a word family?

A **word family** is a group of words that are related and have a similar meaning. Word families are created by adding different prefixes and suffixes to a **root word**.

KEY WORDS
word families
root word

📄 Revise

Here is part of a word family based on the root word **help**: **helpful**, **helpfully**, **unhelpful**, **helpless**. Let's look at some more examples of root words and word families.

Root word	Word family
appear	disappear, reappear, appearance
cycle	tricycle, bicycle, cyclist, recycling
extend	extent, extensive, extension
medicine	medical, medicinal
nature	natural, unnatural, naturally
oppose	opposite, opposed, opposing
possess	possession, possessive
remember	remembrance, remembered

✔ Skills Check

1. Match these words to their word family by writing them in the correct column.

musical   electricity   legality   attentive   irregular

illegal   attention   regulate   electrician   musician

electric	music	attend	regular	legal

Prefixes

What is a prefix?

↺ Recap

A **prefix** is a set of letters added to the beginning of a word to turn it into another word. For example: **un** + happy = **un**happy

re + appear = **re**appear   **dis** + like = **dis**like

📄 Revise

The prefix **in** means the opposite. When a root word begins with certain letters, the prefix **in** changes.

in + active = **in**active   **in** + correct = **in**correct

root word beginning with **m** or **p**: the prefix **in** becomes **im**

im + **m**ature = **im**mature   **im** + **p**ossible = **im**possible ←

ir + **r**egular = **ir**regular ← root word beginning with **r**: the prefix **in** becomes **ir**

il + **l**egal = **il**legal ← root word beginning with **l**: the prefix **in** becomes **il**

The prefix **sub** means under.

sub + zero = **sub**zero

The prefix **inter** means among or between.

inter + national = **inter**national

✔ Skills Check

1. Add a prefix to each word so that the new word has the opposite meaning. Write the new word in the box.

Word	New word
patient	
responsible	
act	
legible	
marine	

💡 Tip

Most prefixes are added to the beginning of root words without any spelling changes.

KEY WORD

prefix

Don't forget that **in** might change to **im** or **ir** or **il**.

Suffixes

What is a suffix?

↺ Recap

A **suffix** is a word ending, or a set of letters added to the end of a word to change its meaning.

Revise

Use the rules in the table to learn how to add suffixes correctly.

Suffix	Rule	Examples
ation	Add to the end of the word. If the root word ends in **e**, remove the **e** before adding **ation**.	inform + ation = inform**ation** ador**e** + ation = ador**ation**
ly	Add to the end of the word. If the root word ends in **le**, change it to **ly**. If the root word ends in **ic**, add **ally**.	sad + ly = sad**ly** gent**le** + ly = gent**ly** bas**ic** + ly = basic**ally**
ous	Add to the end of the word. If the root word ends in **our**, change it to **or** before adding **ous**.	poison + ous = poison**ous** hum**our** + ous = hum**orous**
ion	If the root word ends in **t** or **te** use **tion**.	ac**t** + ion = ac**tion** hesita**te** + ion = hesita**tion**
	If the root word ends in **ss** or **mit**, use **ssion**.	confe**ss** + ion = confe**ssion** per**mit** + ion = permi**ssion**
	If the root word ends in **d**, **de** or sometimes **se**, use **sion**.	exten**d** + ion = exten**sion**, deci**de** + ion = deci**sion** ten**se** + ion = ten**sion**
cian	When the root word ends in **c** or **cs**, use **cian**.	musi**c** + cian = musi**cian**

KEY WORD suffix

✔ Skills Check

1. Change each word below by adding the suffix shown. Write the new word in the box.

Word	Suffix	New word
final	ly	
danger	ous	
politics	cian	

Word	Suffix	New word
dramatic	ly	
prepare	ation	
invent	ion	

Plurals

What is a plural?

↺ Recap

Plural means 'more than one'. **Singular** means 'only one'. There are rules for spelling plural words.

📋 Revise

For most words, to change a word from singular to plural, you add an **s** on the end. However there are a lot of exceptions to this rule! Here are some examples:

Tip

Watch out for the irregular plurals that don't follow the rules!

KEY WORDS

plural
singular
consonants

Add es
when the word ends in **ch**, **sh**, **ss**, **x** or **o**.

Singular	Plural
chur**ch**	chur**ches**
wi**sh**	wi**shes**
dre**ss**	dre**sses**
bo**x**	bo**xes**
potat**o**	potat**oes**

Add es and change y to i
when the word ends in **y** with a **consonant** before it.

Singular	Plural
par**ty**	par**ties**
butter**fly**	butter**flies**

Add es and change f to v
when the word ends in a **f** sound (including **ife**, where the final **e** is not the last sound)

Singular	Plural
hal**f**	hal**ves**
loa**f**	loa**ves**
kn**ife**	kn**ives**

Or
just learn these irregular plurals – they don't follow the rules!

Singular	Plural
tooth	teeth
foot	feet
person	people

✔ Skills Check

1. Write the plural for each singular word below.

Singular	Plural
boy	
curtain	
pony	
ditch	
fish	
sheep	
class	
life	

2. Write the singular for each plural word below.

Singular	Plural
	wishes
	tomatoes
	plates
	feet
	foxes
	loaves
	kisses
	coins

3. Circle the spelling mistake in each sentence below. Then write the correct spelling in the box.

a. George put the knifes and forks on the table ready for lunch.

b. Rakhee watched the butterflys out of the window.

c. The childs went to the theatre to see a funny play.

Longer vowel sounds

↺ Recap

A longer **vowel** sound is a single sound that is longer than the short vowels (**a**, **e**, **i**, **o**, **u**). They are written as a group of letters that contain a vowel. They can be tricky to spell as you can write the same sound in different ways.

📄 Revise

Let's look at some examples of longer vowel sounds and how to spell them.

Vowel sound	Examples
air	fair, hair, pair
ear	bear, pear, wear, tear
are	fare, care, share, scare
ore	more, core, shore, tore
or	short, born, horse, morning
au	autumn, August, dinosaur
aw	crawl, yawn, dawn, draw, saw
ei	vein
eigh	weight, eight, neighbour
ey	they, obey
ough	although
ow	know, snow, grow

Can you think of any more ways to make some of these sounds?

💡 Tip

There are often several different ways to spell the same longer vowel sounds. Try to think of all the possibilities when you are spelling a word!

KEY WORDS

vowel

✔ Skills Check

1. Circle the longer vowel sounds in these sentences.

air ear are	While we waited for my mum to have her hair cut, I shared a pear with my brother. We read a book about a big brown bear. We took it in turns to turn the pages so that it was fair.
ore or au aw	This morning I was drawing a dinosaur when, after a short time, my dad asked if I wanted more breakfast.
ei eight ey	My neighbours who live at number eight have a dog. He is taken for walks around the park. The dog doesn't obey his owners so they keep him on a tight rein.
ough ow	Although it had snowed heavily, school was still open. Later, the caretaker showed us where the melted ice was flowing down the hill.

2. Circle one spelling mistake in each sentence. Then write the correct spelling in the box.

a. In ortumn, the leaves turn to beautiful colours and fall off the trees.

b. In case your feet get wet, could you pack a spare pare of socks, please.

c. I fell off my scooter this morning and my elbow is still very saw.

d. I've had these trousers since I was eit.

e. Althow I am older than my sister, she is taller than me.

Tricky sounds

What do you mean by 'tricky sounds'?

Tricky sounds are letters (or groups of letters) that don't sound the same as they are spelled.

📄 Revise

Here are some tricky sounds and how to spell them.

Words	Spelling	Sound
s**ch**eme, **ch**aracter, e**ch**o	**ch**	k
chauffeur, para**ch**ute, mousta**ch**e	**ch**	sh
di**sc**ipline, fa**sc**inate	**sc**	s
century, **c**entre, **c**ircle, pen**c**il	**c**	s (sometimes called 'soft c')

Look at the tricky sounds highlighted in the examples below.

The noise e**ch**oed around the room.

⬆

Here the **ch** makes a **k** sound.

The para**ch**ute opened in time.

⬆

Here the **ch** makes a **sh** sound.

I was fa**sc**inated by the play.

⬆

Here the **sc** makes a **s** sound.

She drew a **c**ircle with a sharp pen**c**il.

⬆ ⬆

Here the **c** makes a **s** sound.

A **c** before **e**, **i** and **y** often makes a **s** sound. Here are some more examples.

'c' makes a 's' sound before...		
e	i	y
century	**c**ity	**c**ylinder
centre	**c**ircle	**c**ycle
centipede	**c**inema	fan**c**y
cancel	i**c**icle	spi**c**y
i**c**e	pen**c**il	i**c**y

Tip

Words with tricky sounds like these often come originally from other languages, such as French, Greek or Latin. For example:

Spelling	Sound	Origin
ch	k	Greek
ch	sh	French
sc	s	Latin

✔ Skills Check

1. Sort these words according to the sound they make. Write them in the correct column.

came   centre   caterpillar   city   carry   cylinder   cycle   continue   candle   centipede

'k' sound	's' sound

2. Sort these words according to the sound they make. Write them in the correct column.

chemist   machine   chair   brochure   church   chocolate   chef   character   cheese

chorus   chalet   chaos

'ch' sound	'k' sound	'sh' sound

Tricky endings

↻ Recap

Tricky endings are word endings that sound the same, or very similar, but are spelled differently.

It can be easy to make mistakes with these words!

📄 Revise

Let's look at some examples of the tricky endings **sure**, **ture** and **(ch)er**.

Ending	Examples
sure	mea**sure**, trea**sure**, enclo**sure**, plea**sure**
ture	adven**ture**, pic**ture**, furni**ture**, crea**ture**
(ch)er	tea**cher**, ri**cher**, cat**cher**, stret**cher**

Tip 💡

You just have to learn how to spell these words, because you can't always tell the ending from the way they sound!

When you add **er** to a root word ending in **ch**, it sounds the same as (or very similar to) **ture**.

✔ Skills Check

1. Complete the unfinished word in each sentence with the correct ending.

> sure   ture   cher

a. I went on an exciting adven_____ .

b. The tea_____ handed out a sticker at the end of the day.

c. I painted a beautiful pic_____ .

d. The pirate was searching for trea_____ .

e. We went on a na_____ trail to find bugs and worms.

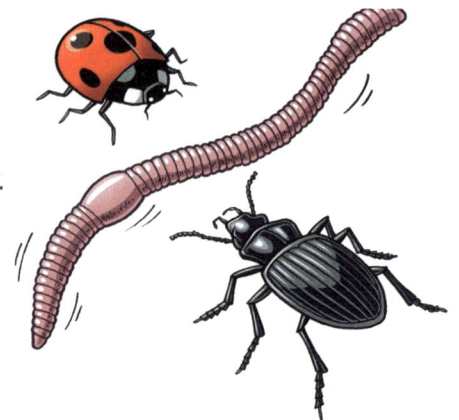

Homophones

What is a homophone?

↺ Recap

Homophones are words that sound the same but are spelled differently and mean different things.

📋 Revise

Here are some examples of common homophones.

How many different examples of homophones can you list?

fair	It's not **fair**. She has more sweets than me.
fare	The bus **fare** was £1.50.

style	He painted a picture in the **style** of a famous artist.
stile	We climbed over the **stile** on our walk in the countryside.

air	Mum opened the window to get some fresh **air**.
heir	The prince is **heir** to the throne.

alter	I need to **alter** my dress so it fits properly.
altar	There are beautiful flowers on the **altar** in church.

aloud	She meant to whisper, but she said it **aloud**.
allowed	I was not **allowed** to go to the party.

KEY WORD

homophones

✔ Skills Check

1. Circle the correct word to complete each sentence. Then write it in the space provided.

a. I had fish and chips for my _____ course at lunchtime. (**main / mane**)

b. She had to decide _____ she wanted blue or green trainers. (**weather / whether**)

c. The princess was _____ to the kingdom. (**air / heir**)

d. You need to _____ the cheese to make cheese scones. (**grate / great**)

e. He decided to _____ the end of the story as it was too sad. (**altar / alter**)

f. We had three turns each on the slide so it was _____ . (**fair / fare**)

Syllables and longer words

↻ Recap

A **syllable** is like a beat in a word. Longer words have more than one syllable. Breaking longer words down into syllables can help you to spell them.

What is a syllable?

📋 Revise

Using syllables is one way to help you remember how to spell longer words. Break the word into parts, say each part slowly and clearly and try to work out how to spell it.

Word	Syllables	Number of syllables
early	ear/ly	2
address	ad/dress	2
exercise	ex/er/cise	3
remember	re/mem/ber	3
separate	sep/a/rate	3
favourite	fav/ou/rite	3
experience	ex/pe/ri/ence	4
material	mat/e/ri/al	4
particular	par/tic/u/lar	4
accidentally	ac/ci/dent/al/ly	5

Clapping a word can help you work out how many syllables it has.

KEY WORD

syllable

💡 Tips

There are lots of ways to learn to spell longer words.

● Practise writing them – again and again!

● Write them in different ways and places. Try chalk outside or coloured icing on a plate. Or write a word in a long line, each time getting smaller and smaller. How many times can you write the word before it gets too small to write? For example:

disappear disappear disappear disappear disappear disappear disappear disappear disappear disappear

● Pick six of your spellings and challenge yourself to use these words in a story.

✔ Skills Check

1. Write the number of syllables for each word below.

Word	Number of syllables
ordinary	
famous	
natural	
perhaps	
interest	
disappear	

2. Try writing some of these longer words. Read the sentence, look at the word, hide it, have a go at spelling it, then check to see if you were right.

Don't forget to cover up the word when you spell it!

Sentence	Word	Spell it
I **promise** I will tidy my room tomorrow.	promise	
You need to **complete** your maths homework by Wednesday.	complete	
There were **various** different cereals to choose from.	various	
Please **continue** with the story.	continue	
Imagine if dinosaurs were still alive today.	imagine	
We had **potatoes** for lunch.	potatoes	

Why not practise spelling some of the words on page 7?

Identifying and summarising main ideas

↺ Recap

Identifying main ideas means reading the text carefully and finding the key points.

Summarising means saying briefly what a passage is about in a few words.

What does identifying and summarising main ideas mean?

目 Revise

Here are some of the main ideas to look for in different types of text.

Non-fiction	Fiction
What the text is about Important facts, taken from across the whole text	The main characters The setting Key events The conclusion

Here is a non-fiction text with the main ideas of each paragraph highlighted. This information is then sumarised below.

Dinosaurs

Dinosaurs lived on the planet millions and millions of years ago, before they mysteriously died out. There are several theories as to why dinosaurs became extinct, but no one really knows.

Scientists learn about dinosaurs by studying their fossils and bones. Scientists who **study dinosaur fossils are called palaeontologists** (pay-lee-on-tol-ogists). There were hundreds of different types of dinosaur.

Tyrannosaurus Rex (**T-rex**) was a **ferocious meat-eating dinosaur**. It had large, pointy teeth to crush through bones as it ate dinosaurs and other animals. It walked around on its two powerful legs and had two small arms. T-rex was an enormous dinosaur.

Triceratops walked on four legs and had a huge **head with three horns on it**. This is how it got its name, as the word *triceratops* means 'three-horned head'. Many scientists believe that the horns were used for protection from the meat-eating dinosaurs. These horns were up to a metre long.

Summary of main ideas

- The text is about dinosaurs.
- Palaeontologists study dinosaur fossils.
- T-rex was a ferocious meat-eating dinosaur.
- Triceratops had three horns on its head.

Tips

- Skim-read the text.
- Highlight main ideas from the text.
- Summarise your main ideas, with a few words or short sentences to explain them.
- If there is more than one paragraph, try to select a main idea from each paragraph.
- Use very few adverbs or adjectives – keep your summary short!

✔ Skills Check

1. a. Read the text below and underline the main ideas.

Reduce, reuse and recycle

A huge amount of waste material is produced in the United Kingdom. Most of this waste is taken to landfill sites and buried in the ground. This can be harmful for the environment.

Recycling means waste products are turned into something new, resulting in less waste ending up in landfill sites. Items that can be recycled include cardboard boxes, yoghurt pots and glass bottles and jars.

Reusing means using items again. For example, plastic bottles can be refilled with water, plastic bags can be used again and again to carry groceries each week, and yoghurt pots can be used as paint pots. Furniture and clothing can be sold or given to someone else who could use them.

Reducing waste involves using fewer materials in the first place. For example, take your own bag when you go shopping, and make sure you use food by its use-by date and don't throw it out unnecessarily.

b. Summarise the main ideas in this text.

_____

_____

_____

_____

_____

Identifying themes and conventions

↻ Recap

What are themes and conventions?

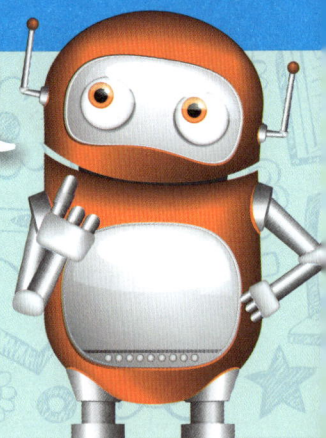

Themes are ideas that go through a text. Conventions are common features that tell you what type of writing it is.

📄 Revise

Let's look at some of the features and conventions of different text types.

Text type	Examples	Conventions and features	Themes
Story	Fairy stories Traditional tales Fables Myths and legends Adventure stories Humorous stories Mystery stories	Characters and setting Beginning, middle, end Problem and solution Paragraphs Sometimes chapters	Good over evil Love and hate/friends and enemies Journeys or quests Wisdom and foolishness Heroism and bravery Morals (in fables) Mythical creatures (in myths)
Poem	Shape poems Limericks Sonnets Ballads Nonsense poems	Verses Capital letter to start each line Exploring and playing with words	Nature and animals Feelings and friendship Everyday happenings Epic tales Historical events
Recount	Diaries Newspaper articles Historical recounts	Past tense In time order Pictures/photos and captions	Personal life events Journeys and holidays News stories Historical events
Report	Information texts Explanation	Headings Paragraphs Pictures/diagrams/photographs Bullet points	Factual subjects (such as geography, science, history) Accurate/reliable information
Instructions	Recipes How to make…	Equipment list Numbered points Commands (put, chop, mix) Conjunctions of time (first, then)	Details of how to make things (such as food, crafts, toys, furniture)

✔ Skills Check

1. Read this text, then answer the questions below.

Theseus and the Minotaur

Theseus the Prince of Athens set sail on a warm, windy afternoon. He was sailing for the island of Crete. Theseus had bravely volunteered as one of the seven young men to be fed to the Minotaur every year. His quest was to defeat the furious, hungry Minotaur and free the people of Athens.

After a long journey, Theseus finally arrived at the island. He was taken to the complex maze beneath the castle of King Minos, where the Minotaur lay waiting for his meal. While Theseus fearlessly prepared for his mission, a princess appeared and gave him a ball of string.

Theseus used the ball of string to lay a trail behind him as he walked through the confusing maze of tunnels. He could smell the horrid stink of the Minotaur, an evil creature with the head of a bull and the body of a man.

Theseus heroically killed the creature and then followed his trail of string back to the entrance.

a. What type of text is this? Tick **one**.

Instructions ☐   Myth ☐   Poem ☐

b. What is the creature called? _____

c. Who is the hero? _____

d. What is the hero's aim?

_____

_____

e. Which theme can you see in this text? Tick **one**.

Journeys or quests ☐   Wisdom and foolishness ☐

Love and hate ☐

Retrieving and recording information

What does retrieving and recording mean?

↺ Recap

Retrieving information means finding the information you need from a text to answer questions.

Recording information means writing it down.

📋 Revise

See if you can spot the answers in the text.

Look at the example questions below. You can look for key words in the passage that are in the question to help you, they have been highlighted below.

1. Read this text, then answer the questions below.

Milk

Milk can be produced by cows or goats. It is a highly nutritious white drink and is also used to produce lots of different foods, such as cheese, butter and yoghurt. These foods are called dairy products.

Dairy farmers milk cows using a milking machine. The milk is then cooled in large tanks to keep it fresh. Often, the milk is collected in special vehicles called milk tankers which take the milk to a large dairy. At the dairy the milk is treated to kill any bad bacteria and to make sure it is safe to drink. This is called pasteurisation.

Milk contains calcium which is important for helping our bodies to develop strong, healthy bones and teeth. It also contains other vitamins that help us to grow and stay healthy. Milk can be enjoyed as part of a healthy diet on its own or in other foods.

a. Which food product is created from milk? Tick **one**.

Lettuce ☐   **Cheese** ✓   Crisps ☐

b. What name is given to the foods produced from milk?

These foods are called dairy products.

c. What is pasteurisation?

Pasteurisation is when milk is treated to kill any bad bacteria.

d. Why is calcium good for our bodies?

Calcium helps our bodies to develop strong, healthy bones and teeth.

✔ Skills Check

1. Read this text, retrieve the information you need and record your answers below.

Rocks

There are three main types of rock. These are igneous, metamorphic and sedimentary rocks.

Igneous rocks are formed when magma is forced from the earth in a volcano. As the magma cools, it forms rocks. Examples include basalt and granite. Granite is very practical and has many uses, including paving stones, kitchen worktops and gravestones.

Sedimentary rocks are formed from sediments such as other tiny rocks and animal skeletons pressed together in layers. Examples include chalk, clay and sandstone. These are softer rocks. Chalk can be used to draw with.

Metamorphic rocks are rocks that have changed from another type of rock due to heat and pressure. Examples include slate and marble. Slate is often used for roof tiles.

Now read the questions and have a go at finding the answers in the text.

a. How many different types of rock are there? _____

b. What can granite be used for? Give two examples from the text.

_____

c. Which rock can be used to draw with? Tick **one**.

Marble ☐   Basalt ☐   Chalk ☐

d. How are metamorphic rocks created?

_____

Making predictions

How do you make predictions?

Making predictions means saying what you think will happen next using evidence from the text.

📄 Revise

Read this text. What happens next? Some clues have been highlighted that help you predict.

Megan's friend Noah had come round to her house after school.

"Wow! Can we play with that?" asked Noah, pointing to a small remote control car up on the shelf.

"Oh, **I'm not sure**, Noah. **It's really fragile** and **it was my dad's when he was a little boy**," explained Megan, **looking worried**.

"It's OK, I'll be careful," said Noah as he pulled it down off the shelf. **Tossing the car onto the floor**, he started pressing buttons and pulling levers on the controller. The little car was sent **whizzing off towards the wall**.

"No! Gently! One lever at a time," shouted Megan as she watched the old toy car **spinning round in circles**. Then suddenly…

What do the clues tell us?

"I'm not sure"/ looking worried these suggest that Megan thinks it is not a good idea, and something bad could happen.

"It's really fragile… it was my dad's when he was a little boy" shows that it could break easily, it belongs to someone else, is old and perhaps precious.

Tossing the car onto the floor suggests that Noah is not being careful at all.

Whizzing off towards the wall/spinning round in circles implies that the car is out of control.

So the clues suggest that the toy car gets damaged or broken in some way. Now we can use these clues to make a prediction, give a reason for it and explain the evidence.

prediction →

explanation of clues in the text →

← reason

I think that the toy remote control car gets badly damaged because Noah is being careless with it and the toy goes out of control. The text says Noah tossed the car onto the floor and that it was whizzing and spinning round. Megan is also worried which suggests something bad could happen.

Tips

- Underline clues in the text.
- Give a reason for your prediction.
- Explain what the clues are telling you.

✔ Skills Check

1. Read this text, then answer the questions below.

A new bike

Ollie had a new bike for his birthday. It was shiny and blue and he was really proud of it. That morning, his mum asked him to go to the shop and buy some bread, as they had run out.

Ollie cycled past two older boys sitting on the wall opposite the shop.

"Nice bike," one of the boys whispered to the other and nodded in Ollie's direction. They looked like they were up to something and made Ollie feel nervous.

Outside the shop Ollie looked in his rucksack for his bike lock, but he had forgotten it. At first, he couldn't decide what to do. His mum would be cross if he didn't get the bread and you weren't allowed to take bikes into the shop, so he left his bike and ran into the shop.

When Ollie came out of the shop, the boys across the road had gone and so had his bike.

a. What do you think happened to Ollie's bike?

_____

_____

b. Use evidence from the text to explain your answer.

_____

_____

_____

_____

Making inferences

↻ Recap

How do you make inferences?

Sometimes an author doesn't tell you everything in a text. You have to use clues in the text to work out what is happening. This is called making inferences.

Revise

In this example, the author does not simply write: 'Oliver was nervous about going horse riding'. Instead, she suggests this, by describing Oliver's behaviour.

evidence that suggests Oliver might be feeling nervous

Oliver was at the stables. He was about to go for his first riding lesson **since he had fallen from a horse**. He was wearing his riding hat and riding boots. Oliver watched as Maya put the saddle on **the huge horse**. As he watched, his **hands got all sweaty**, his **face went pale** and his **eyes grew wider**. Oliver **stepped slowly back, away from the horse**.

"Come on Oliver, on you get," said Maya cheerfully.

Oliver didn't move.

Now you can use the evidence in the text to explain what you have inferred about Oliver's behaviour. For example:

Oliver's hands got all sweaty, his face went pale and his eyes grew wider suggests that he is feeling nervous and scared.

The horse was huge and Oliver had not been riding since he fell off are reasons *why* he might be scared.

Oliver backed away from the horse and wouldn't get on when Maya asked him to shows that he didn't want to go.

This is fun – it's just like solving a puzzle!

Tips

- Read the text carefully.
- Highlight the evidence (clues) in the text.
- Use the evidence from the text to support and explain your answer.

It can be helpful to underline the clues you find in the text.

✔ Skills check

1. Read this text, then answer the questions below.

Farah sat at the dinner table. Everyone else had finished eating. Farah's plate sat in front of her. She had eaten everything *except* the peas. Farah stabbed a pea onto the end of her fork and slowly lifted it to her mouth. She made a face before the pea even touched her lips.

Do you think Farah likes peas? Circle **one**.

Yes   No

Explain your answer, using evidence from the text.

_____

_____

2. Read this text, then answer the questions below.

Ali was going to a music concert and had been waiting for the tickets to arrive. He ran to the door when he heard the post drop through the letter box and onto the floor. Picking up the post Ali searched through the letters until he found them! His tickets had finally arrived. Ali ripped open the envelope and pulled out the colourful music concert tickets. A huge smile spread across his face.

How is Ali feeling about going to the music concert?

_____

Explain how you know.

_____

_____

_____

Remember to use the evidence from the text to explain your answer.

Language features

What are language features?

Language features are things that writers use to make their text more interesting.

Revise

This table summarises some important language features to look out for.

Language feature	Description	Examples	The effects are to…
Choice of words	Interesting words: • adjectives • adverbs	**delicious** cake **cute**, **furry** mouse **Suddenly**, the door banged open.	• make a text more interesting • create a picture for the reader
Rhyming words	When word endings sound the same	m**ap**, t**ap**, cl**ap**, z**ap**, d**ate**, m**ate**, cr**ate**, l**ate**	• make a text easier to remember and fun to read aloud • make phrases stand out • add rhythm/beat and help it to flow
Repeated sounds	When words begin with the same letter or sound	**S**lowly, the **s**nail **s**lithered across the path.	
Words that describe sounds	When words sound like the sounds they describe	**bang**, **beep**, **boom**, **clap**, **clang**, **clip-clop**, **flap**, **moo**, **pop**, **splash**, **spit**, **toot**, **zip**	• create a sound image of the setting/event and bring it to life

Tip

Look for different language features in a text and think about *why* the author used that feature. This will help you to explain what effect the feature has on the text.

✔ Skills Check

1. Read this text, then answer the questions below.

It was the middle of the day in the school holidays. Jane slept peacefully until her mother pulled back the curtains and a sea of light flooded the room. Slowly, Jane sat up and stretched. "Was it really time to get up already?" she asked herself. A bee buzzed noisily around the room. Once she was dressed, Jane poured herself a glass of fabulously fresh fruit juice.

a. Find a word that describes a sound in the picture.

_____

b. What effect does this have?

_____

c. Find an example of repeated sounds in the text.

_____

d. Which adverbs are used in the text? Write them below.

_____

e. What does 'a sea of light flooded the room' mean?

_____

_____

Words in context

What are words in context?

↻ Recap

Sometimes when you are reading, you come across a word that you don't understand. You need to use the clues in the text (the context) to work out the meaning.

📄 Revise

Some words have more than one meaning and you need to work out which meaning is meant.

Word	Meaning	Alternative meaning
row	line of something	to row a boat
light	opposite of dark	not very heavy
trip	journey	to fall over
duck	type of bird	to bend/lower your head
squash	fruit drink made with water	to squeeze/crush
date	dried fruit	day on a calendar
lift	elevator	to pick up
flat	somewhere to live	level/smooth

Some words even have three or more meanings! For example, squash is also a ball game played with rackets.

Some words you might not know at all and you need to work it out using the text.

1. What does the word 'harmoniously' mean in this sentence? Tick one.

Owen and Evie went to the sports centre. They hired rackets and a ball. They played squash **harmoniously**. They had lots of fun.

happily ✔   heroically ☐   nastily ☐

In this case, you should tick happily as it links best with 'fun' in the next sentence.

✔ Skills Check

1. **What does the word 'flat' mean in this sentence? Tick one.**

> We live in a flat in the centre of town.

Level ☐   Somewhere to live ☐   Smooth ☐

2. **Which word has a similar meaning to 'swung' in this sentence? Tick one.**

> I stood back as my brother swung the golf club at the little white ball.

waved ☐

moved ☐

bashed ☐

3. **a.** What does the word 'expedition' mean in this text? Tick **one**.

> At the weekend, I went on an expedition to the seaside.
> We travelled for several hours down the motorway.
> Finally, we arrived at the seaside where we had ice
> cream and built sandcastles.

memory ☐   journey ☐   entrance ☐

b. What evidence in the text tells you this?

_____

_____

Presentational features: non-fiction

↻ Recap

What are presentational features in non-fiction texts?

Texts can be presented in different ways, called presentational features. They give clues about what type of text it is, highlight important parts of the text and even affect its meaning.

📄 Revise

Here are some examples of presentational features in non-fiction texts.

Feature	Purpose	Effect
Bold writing	To highlight words and headings	They may change the emphasis the reader puts on these words in a sentence. When used in headings, they help the reader know what the text is about.
Italics	To add emphasis or show that something is important	
Underlining	To make a title, heading or subheading stand out, or to emphasise a word	
Headings/ subheadings	To label paragraphs or sections of text	They help direct the reader to the right section of the text.
Paragraphs	To break the text into sections	They make the text easier to read and understand by organising the content into sentences linked by a theme.
• Bullet points or 1. Numbered points	To summarise the text or to order parts of it	These help the author to highlight the key points/facts for the reader. Numbered points tell the reader in what order to complete a task, which may be important to get the best results.
Pictures/ photographs/ diagrams	To show what something looks like, or give visual clues	Images give the reader a clear visual picture to add to the writing. Captions tell the reader what the picture is for and what it shows
Captions	To label pictures or photographs	
Layout	To enhance the effectiveness of the text. For example: newspaper articles – in columns, large headings instructions – numbered points	This helps the reader to identify the text type and purpose, and to get the most out of the text.

✔ Skills check

1. Read this text, then answer the questions below.

1 [ ]

→ ## Fruit kebabs

2 [ ]

→ ### Ingredients
- 200g strawberries
- pineapple
- large bunch of green grapes

3 [ ]

→ - banana
- 100g milk chocolate
- 12 wooden kebab sticks

4 [ ]

Fantastic fresh
fruity kebabs

5 [ ]

Method

1. First, wash the strawberries and remove the stalks.

2. *Carefully*, use a sharp knife to cut the pineapple into 2cm cubes.

3. Peel and slice the banana.

4. Then thread the fruit onto the kebab sticks, one at a time. Use about 6 pieces of fruit on each stick, and try to create a repeating pattern with the fruit you choose.

5. After that, slowly melt the chocolate in a microwave.

6. Dip the fruit kebabs in the melted chocolate and then place them on greaseproof paper on a tray.

7. Finally, put the tray of kebabs in the fridge for about 30 minutes to allow the chocolate to set.

a. Use the features below to write labels in the numbered boxes around the text.

subheading   heading   bullet points   caption   picture

b. Why is the word '*Carefully*' in italics?

_____

c. Why are the sentences in the **Method** numbered?

_____

Presentational features: fiction

↻ Recap

What are presentational features in fiction texts?

Fictional texts (such as poetry, stories and comic strips) have their own presentational features. Poetry in particular uses a wide variety of features to help the reader explore language on the page.

🗎 Revise

Here are some examples of presentational features in fiction texts.

Text type	Feature		Effect
Story	Paragraphs	These organise the story into sections	This makes it easier for the reader to follow the text.
Poetry	Font style	This changes the shape and appearance of the letters	These can be used to emphasise the meaning of a word. For example: *wobble* The baby was really tiny.
	Font size	Individual words/phrases can be bigger or smaller	
	CAPITAL LETTERS	These show that words are loud or being shouted	The reader can shout when reading aloud to add surprise. For example: "BOOM went the big bass drum!"
	Shape	Poems can be written in different shapes around or across the page	This can add meaning and interest to a poem. For example, a poem about a tower might be written as a vertical list.
	Length of lines	Lines in a poem can be very short or very long	This changes the flow of a poem and can make it short, snappy and fast, or slow and gentle.
Playscripts	Bold text	The character names are often bold and spaced out from the spoken text	Makes it easier for the reader to identify the characters.
	Italic text	Stage directions are often in italics to show what is happening	Tells the reader what the characters are doing to give them direction.

Tips

Presentational features can make the text look fun and exciting!

The layout, presentational and language features of a text can add meaning and emphasise text, as well as adding interest to the look of the text on the page.

✔ Skills Check

1. Read this poem, then answer the questions below.

Long, extended, lengthy snake. Slithering, sliding, gliding snake. Powerful, strong, deadly snake <

a. Why do you think the poet has written the poem like this?

_____

_____

b. How has the poet used presentational features to create the snake's head?

_____

_____

2. Read this playscript, then answer the questions below.

Red Riding Hood is skipping through the wood holding a basket and bumps into the wolf.
Red Riding Hood: Hello Mr Wolf.
Wolf: Why, hello little girl. What are you doing on this fine day?
Red Riding Hood: I'm off to see my grandmother to give her this basket of food.

a. Which presentational features are used in this text? Tick **two**.

Bold writing ☐   Speech bubbles ☐   Italics ☐   Subheadings ☐

b. How does the stage direction help the reader to understand the text?

_____

_____

Maths
Made Simple
Ages 8–9

Numbers to 9999

The value of a digit depends on which column it is in.

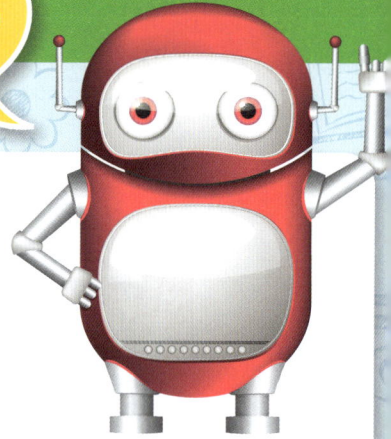

↻ Recap

Our number system uses 100s, 10s and 1s.
265 in words is two hundred and sixty-five.

100s	10s	1s
2	6	5

The **place value** of the digit 2 is 100s. The digit 2 represents 200.
The **place value** of the digit 6 is 10s. The digit 6 represents 60.
The **place value** of the digit 5 is 1s. The digit 5 represents 5.

▤ Revise

This number is five thousand, seven hundred and nine.

1000s	100s	10s	1s
5	7	0	9

Zeros are important. They help to show the place value of all the digits.

We can write numbers in words or using numerals.

- six thousand, four hundred and eighty-five ⟶ 6485
- three thousand, nine hundred and one ⟶ 3901
- two thousand and seven ⟶ 2007

💡 Tips

- When you're asked to write a number, write the place values above the digits if you're stuck.
  Say this number. 8704

  Write the place values above the digits.

1000s	100s	10s	1s
8	7	0	4

  The number is eight thousand, seven hundred and four.

DID YOU KNOW?

A thousand is ten hundreds.

Talk maths

What are the biggest and smallest numbers you can make?

| 0 | 1 | 2 | 3 | 4 | 5 | 6 | 7 | 8 | 9 |

Write the digits 0 to 9 on some pieces of card or paper. Use them to make ten different 4-digit numbers. Write them down, and then read them aloud.

✔ Check

1. Write these numbers in words.

   a. 7380 _____

   b. 2069 _____

2. Write these numbers in digits.

   a. six thousand, eight hundred and forty-one _____

   b. five thousand and two _____

3. Arrange these numbers in order, from smallest to largest.

   1612   5000   725   8   250   3875   92   9999

   _____ _____ _____ _____ _____ _____ _____ _____

4. Complete this chart.

1000 more	3350				
Number	2350	1243	4789	7000	8999
1000 less	1350				

⚠ Problems

Blinkton	Dipton	Mumsford	Pilbery	Wester
4307	974	3824	1092	1003

Five villages count their populations.

Brain-teaser a. Which village has the smallest population? _____

b. Which village has the largest population? _____

Brain-buster The number of people living in all the villages, added together, is 11,200.

Write this in words. _____

73

Estimating and rounding

↺ Recap

We sometimes have to round numbers. This can help us to estimate amounts and calculations more easily.
To round a number to the nearest 10 we look at its position on the number line.

We then look for the nearest 10.
- 12 rounds down to 10.
- 18 rounds up to 20.
- 15 is halfway, but we always round it up.

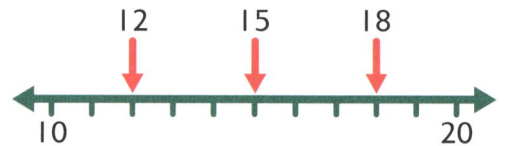

📋 Revise

We can also round large numbers to the nearest 100 or 1000.

432 rounds down to 400.

450 and 473 both round up to 500.

1268 rounds down to 1000.

1500 and 1647 both round up to 2000.

💡 Tips

We can round numbers to give a quick estimate. This is useful for seeing if your answers are about right.
So, for 372 + 221, a quick estimate is 400 + 200 = 600.

- Think carefully about what you want to round to: 10s, 100s or 1000s.
  - 6852 rounds to the nearest 10 as 6850.
  - 6852 rounds to the nearest 100 as 6900.
  - 6852 rounds to the nearest 1000 as 7000.

🗨 *Talk maths*

What is 45 rounded to the nearest 10?

Write down six different numbers between 0 and 9999. For example:

| 45 | 94 | 143 | 2530 | 5265 | 9250 |

Work with a partner and challenge each other to round the numbers.

What is 2530 rounded to the nearest 1000?

✔ Check

1. Round each number to the nearest 10, 100 and 1000.

	To the nearest 10	To the nearest 100	To the nearest 1000
a. 77			
b. 583			
c. 1232			
d. 3765			

2. Estimate to the nearest 10. Round each number before you add.

   a. 43 + 36 _____   **b.** 25 + 36 _____   **c.** 82 + 37 _____

3. Estimate to the nearest 100. Round each number before you add.

   a. 423 + 186 _____   **b.** 75 + 215 _____   **c.** 452 + 821 _____

4. Estimate to the nearest 1000. Round each number before you add.

   a. 4233 + 1836 _____   **b.** 825 + 3336 _____   **c.** 7852 + 3500 _____

⚠ Problems

Blinkton	Dipton	Mumsford	Pilbery	Wester
4307	974	3824	1092	1003

Brain-teaser
Look at these different village populations.
Which villages have populations that are 4000, when rounded to the nearest 1000?

_____

Brain-buster Estimate the total number of people living in all five villages to the nearest 1000.

_____

Counting in steps

1	2	3	4	5	6	7	8	9	10
11	12	13	14	15	16	17	18	19	20
21	22	23	24	25	26	27	28	29	30
31	32	33	34	35	36	37	38	39	40
41	42	43	44	45	46	47	48	49	50
51	52	53	54	55	56	57	58	59	60
61	62	63	64	65	66	67	68	69	70
71	72	73	74	75	76	77	78	79	80
81	82	83	84	85	86	87	88	89	90
91	92	93	94	95	96	97	98	99	100

↺ Recap

When we count in steps, we add or subtract the same number each time. In this 100-square all the multiples of 4 have been shaded.

Can you count in steps of 5? Shade each multiple of 5 the same colour.

Which numbers are multiples of 4 **and** 5?

📄 Revise

Use the 100-square above to count in steps of 6, 7 and 9.

0	6	12	18	24	30	36	42	48…
0	7	14	21	28	35	42	49	56…
0	9	18	27	36	45	54	63	72…

Try colouring in these steps on the 100-square.

You can count in steps of any number. You need to learn to do this with 25s and 1000s.

0	25	50	75	100	125	150
0	1000	2000	3000	4000	5000	

Can you see any patterns?

Can you keep these sequences going?

Remember that counting in steps and times tables facts have lots in common.

💡 Tips

- 6 + 6 + 6 is 'three lots of 6', or 3 × 6. They all equal 18.
- 7 + 7 + 7 + 7 + 7 + 7 + 7 + 7 + 7 is 'nine lots of 7', or 9 × 7. They all equal 63.
- 9 + 9 + 9 + 9 + 9 is 'five lots of 9', or 5 × 9. They all equal 45.

Talk maths

Try racing against the clock. What is the fastest you can count aloud in 6s, 7s and 9s up to 100 without making a mistake?

What about counting in 25s up to 300?

Or counting in 1000s up to 10,000?

✔ Check

1. **Complete these sequences.**

   a. Count on in steps of 6.   **36**, _____ , _____ , _____ , _____

   b. Count on in steps of 7.   **56**, _____ , _____ , _____ , _____

   c. Count on in steps of 9.   **45**, _____ , _____ , _____ , _____

   d. Count on in steps of 25.   **350**, _____ , _____ , _____ , _____

   e. Count on in steps of 1000.   **2000**, _____ , _____ , _____ , _____

2. **Complete these sequences.**

   a. Count back in steps of 6.   **90, 84**, _____ , _____ , _____ , _____

   b. Count back in steps of 7.   **77, 70**, _____ , _____ , _____ , _____

   c. Count back in steps of 9.   **81, 72**, _____ , _____ , _____ , _____

   d. Count back in steps of 25.   **875**, _____ , _____ , _____ , _____

   e. Count back in steps of 1000.   **9000**, _____ , _____ , _____ , _____

⚠ Problems

Brain-teaser Joe saves £6 a week for eight weeks. Kate saves £9 a week for five weeks. Who has the most money, and how much more do they have?

_____

Brain-buster Kate's older sister wants to save £800 for a holiday abroad. If she can save £25 a week, how long will she have to save for?

_____

Negative numbers

↻ Recap

$$3 + 2 = 5 \qquad 9 - 3 = 6$$

0 1 2 3 4 5 6 7 8 9 10

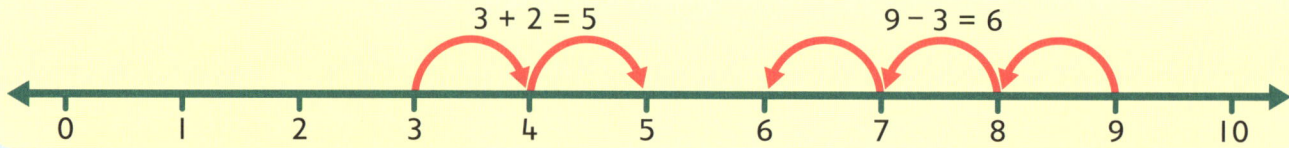

When we **count on** we move to the right along the number line.
$3 + 2 = 5$

When we **count back** we move to the left.
$9 - 3 = 6$

📄 Revise

Numbers can be negative as well as positive.

−5 −4 −3 −2 −1 0 1 2 3 4 5

Look at the numbers on each side of zero.
On the right, the more you move away from zero, the bigger the numbers get (5 is bigger than 2).
On the left, the more you move away from zero, the smaller the numbers get (−2 is bigger than −5).

We can count back through zero. Use the number line to count back in steps of 1.

Start at 1 and count back 2. You should stop at −1.
Start at 2 and count back 4. You should stop at −2.
Start at 4 and count back 7. You should stop at −3.
Start at 1 and count back 5. You should stop at −4.
Start at 3 and count back 8. You should stop at −5.

DID YOU KNOW?

°C means degrees Celsius. Zero degrees Celsius is the temperature at which water freezes.

💡 Tips

Notice that when you count in steps below zero, you still count in order.

- Temperature is a great way to practise counting back through zero.

If you start at 5°C and count back 5°C, you end at 0°C.

If you start at 2°C and count back 10°C, you stop at −8°C.

💬 Talk maths

Draw a thermometer, as large as you can, and mark its scale from –10°C to 10°C. Ask someone to check that you have written the scale correctly.

Working with someone, place a small object or counter somewhere on the scale of the thermometer. Challenge your partner to count on or back.

If you want to make this harder, don't mention the temperature and just present your challenges as subtractions.

Count back 7 from 3°C.

3°C, count back 7, stops at –4°C.

✔ Check

–10 –9 –8 –7 –6 –5 –4 –3 –2 –1 0 1 2 3 4 5 6 7 8 9 10

1. Use the number line to help you solve these problems.

   a. 6 count back 5 = _____

   b. 3 count back 4 = _____

   c. 1 count back 4 = _____

   d. 2 count back 7 = _____

   e. 0 count back 5 = _____

   f. 5 count back 10 = _____

   g. 8 count back 9 = _____

   h. 10 count back 16 = _____

2. Say how many steps have been counted back on each number line.

   a.

   –5 –4 –3 –2 –1 0 1 2 3 4 5 _____

   b.

   –5 –4 –3 –2 –1 0 1 2 3 4 5 _____

   c.

   –5 –4 –3 –2 –1 0 1 2 3 4 5 _____

⚠ Problems

Brain-teaser The temperature starts at 2°C, and then goes down 3°C. What is the new temperature? _____

Brain-buster The temperature drops from 9°C to –6°C. How many degrees has it dropped by? _____

Roman numerals

↻ Recap

In our number system, all of our numbers are made using ten different digits.

0 1 2 3 4 5 6 7 8 9

Using place value (100s, 10s and 1s), we can use these ten digits to represent any number we wish.

📄 Revise

There are many different number systems. One other that we still use is Roman numerals. The Romans used letters to represent some numbers.

There are five Roman numerals that you need to know.

I (1)   **V** (5)   **X** (10)   **L** (50)   **C** (100)

By using these numbers together, they could make any number, but it isn't always easy! The chart below can help you to learn Roman numerals to 100.

Number	1	2	3	4	5	6	7	8	9	10
Roman numeral	I	II	III	IV	V	VI	VII	VIII	IX	X
Number	11	12	13	14	15	16	17	18	19	20
Roman numeral	XI	XII	XIII	XIV	XV	XVI	XVII	XVIII	XIX	XX
Number	30	40	50	60	70	80	90	100		
Roman numeral	XXX	XL	L	LX	LXX	LXXX	XC	C		

💡 Tips

XCIV plus LIX equals...

- It all seems very complicated, but if you learn some important numbers it can be okay.
- Learn the Roman numerals 1 to 10 by heart. It will make other numbers easier to understand.
- Also, pay close attention to how they make 4 (IV), 9 (IX), 40 (XL) and 90 (CX).

Talk maths

Challenge each other to write your ages in Roman numerals.

I   I   I   V   X   X   X   L   C

Make a set of Roman numeral cards. It is important to have three 'one' cards, and three 'ten' cards. These can make any number between 1 and 100.

Next, challenge a friend to make numbers. You can either call out a number, such as seventy-three, or arrange some Roman numerals.

Whenever anyone gives an answer, they must explain their working out.

✔ Check

1. Change these numbers to Roman numerals.

    a. 4 _____   **b.** 11 _____   **c.** 25 _____   **d.** 19 _____

    e. 52 _____   **f.** 45 _____   **g.** 90 _____   **h.** 87 _____

2. Change these Roman numerals to numbers.

    a. VI _____   **b.** IX _____   **c.** XVII _____   **d.** XXII _____

    e. LV _____   **f.** XL _____   **g.** LXXXVIII _____   **h.** XC _____

3. Which digit can't be shown with Roman numerals?

    _____

⚠ Problems

Change these Roman numerals into numbers, and then complete the calculations.

Brain-teaser XIV plus XXIII

Brain-buster XCI minus LXV

Mental methods for addition and subtraction

> And, you can use these facts to do harder mental calculations. 70 + 80 = 150, 700 + 800 = 1500

↻ Recap

You will probably know how to do some calculations in your head, and also how to do them by writing on paper. You must know your number bonds. Each number bond gives you four different facts.

Look at these.

| 7 + 8 = 15 | 8 + 7 = 15 | 15 − 8 = 7 | 15 − 7 = 8 |

📄 Revise

> And this means that 12 + 25 = 37, 37 − 25 = 12 and 37 − 12 = 25.

Here are two very useful mental methods.

Partitioning a number into 10s and 1s is useful: 25 + 12 = 25 + 10 + 2 = 35 + 2 = 37

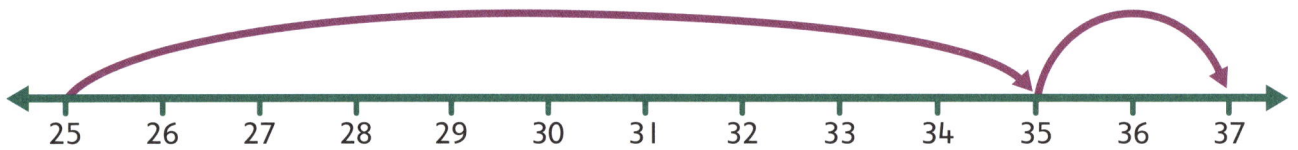

Bridging is used when one number is close to a 10.
For example, for 75 + 98, instead of adding 98, add 100 and then subtract 2.

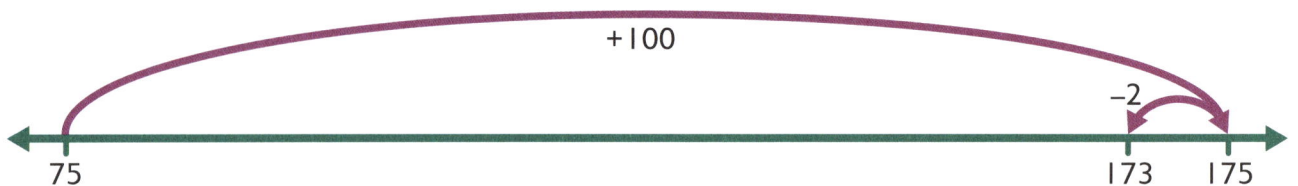

You can use both of these methods for subtraction too.

| 38 − 13 = 38 − 10 − 3 = 25 | 46 − 9 = 46 − 10 + 1 = 37 |
| 57 − 24 = 57 − 20 − 4 = 33 | 246 − 97 = 246 − 100 + 3 = 149 |

> So, 75 + 98 = 75 + 100 − 2 = 173.

💡 Tips

> Remember, if in doubt, write it out!

- Look carefully at calculations before you try to solve them in your head – you might prefer to solve them with a written calculation. Look at the calculations below, which ones would you solve mentally?

| 25 + 96 | 645 + 377 | 89 − 30 | 204 − 157 |

Talk maths

Demonstrate to an adult how to use different mental methods for addition and subtraction. Remind them that the first thing to decide is whether a mental method is appropriate or not.

Use the calculations in the box to get you started.

Calculation	OK for mental methods?
23 + 11	yes
47 + 98	yes
645 + 123	yes
472 + 687	maybe not!
35 – 9	yes
642 – 102	yes
743 – 510	yes
403 – 187	maybe not!

✔ Check

1. Add these numbers, using mental methods.
   a. 36 + 14 = ____   b. 64 + 25 = ____   c. 95 + 13 = ____   d. 343 + 51 = ____
   e. 12 + 11 = ____   f. 67 + 19 = ____   g. 67 + 28 = ____   h. 467 + 97 = ____

2. Subtract these numbers, using mental methods.
   a. 27 – 13 = ____   b. 64 – 22 = ____   c. 90 – 46 = ____   d. 357 – 206 = ____
   e. 45 – 9 = ____   f. 72 – 11 = ____   g. 65 – 19 = ____   h. 436 – 198 = ____

3. Solve these calculations using mental methods.
   a. 43 + 34 = ____   b. 67 + 32 = ____   c. 335 + 150 = ____   d. 2231 + 3607 = ____
   e. 86 – 24 = ____   f. 54 – 33 = ____   g. 276 – 143 = ____   h. 5493 – 3170 = ____

⚠ Problems

Brain-teaser

a. Tina has £47 in her bank account, and she receives another £52 for her birthday.
   How much does she have altogether? _____

b. James is travelling from London to Newcastle. The distance is 295 miles.
   So far he has travelled 190 miles. How far does he have to go? _____

Brain-buster

Joe's mum buys a new car. The car costs £4500 and she pays a deposit of £2200.

How much does she still have to pay? _____

Written methods for addition

↻ Recap

There are formal written methods for adding numbers. You may have been taught methods a bit different to this one. You should use whichever method you are comfortable with.

We arrange the numbers so that the place value of their digits line up.

	100s	10s	1s
	1	4	5
+	3	2	7
	4	7	2
			1

That's why some people call it column addition. The secret is to add the digits in each column like 1s. Remember to add on any numbers that have been carried.

📄 Revise

The biggest digit we can write in a column is 9. If the digits in the 1s column add up to more than 10, we must carry the 10 to the next column, and leave the 1s behind. You can do the same when adding digits in the other columns.

	2	4	6	1
+	1	5	7	8
	4	0	3	9
		1	1	

Sometimes the digits add up exactly to 10. In this calculation, we carry one 10 an leave no 1s behind. In the 100s column, we carry one 1000 and leave no 100s behind.

	3	2	7	4
+	2	8	1	6
	6	0	9	0
		1		1

💡 Tips

- Remember to carry any 10s, 100s or 1000s to the next column.

You can add as many numbers as you want using column addition.

	1	3	5	2
	2	8	1	5
		6	3	4
+	3	0	2	5
	7	8	2	6
		1	1	1

💬 Talk maths

Look at this addition and explain it aloud. Say what is happening at each stage. Make sure you work in the correct order, right to left.

```
    4   8   6   7
+   2   5   0   8
─────────────────
    7   3   7   5
```

Try some column additions using three or four numbers, explaining each stage to someone you know.

✔ Check

1. Complete each of these additions.

a.
```
    2   3   7
+   4   8   1
─────────────
```

b.
```
    7   4   6
+   2   5   7
─────────────
```

c.
```
    2   4   2   6
+   1   4   8   5
─────────────────
```

d.
```
    1   3   2   7
    2   4   5   1
+   3   8   0   6
─────────────────
```

2. Complete these additions. Use a written method on squared paper.

a. 365 + 237   b. 467 + 259   c. 4459 + 3557   d. 2406 + 2205 + 3670 + 1338

⚠ Problems

Brain-teaser

This chart shows the number of people attending three village fetes.

Village	Plink	Plank	Plonk
Number of visitors	825	936	89

What is the total number of visitors to the three fetes? _____

Brain-buster

This chart shows the number of fans at three rock concerts.

Band	Crush	Push	Mush
Fans	6455	7106	3453

What is the total number of fans at the three concerts? _____

85

Written methods for subtraction

↺ Recap

There are formal written methods for subtracting numbers. You may have been taught methods a bit different from this one. You should use whichever method you are comfortable with.

$$
\begin{array}{r}
{}^5\!\!\not{6} \quad {}^{13}\!\!\not{4} \quad {}^1 5 \\
-\quad 2 \quad\ 7 \quad\ 8 \\
\hline
3 \quad\ 6 \quad\ 7 \\
\hline
\end{array}
$$

DID YOU KNOW?

Subtraction is the *inverse* of addition. You can use addition to check your subtractions.

73 – 48 = 25   checking…

25 + 48 = 73   correct! ☺

Notice how we exchange 100s for 10s and 10s for 1s.

🗎 Revise

Just like with addition, we can use the place-value columns to subtract larger numbers.

$$
\begin{array}{r}
{}^5\!\!\not{6} \quad {}^1 2 \quad {}^4\!\!\not{5} \quad {}^1 4 \\
-\quad 1 \quad\ 7 \quad\ 3 \quad\ 8 \\
\hline
4 \quad\ 5 \quad\ 1 \quad\ 6 \\
\hline
\end{array}
$$

You need to be very careful at each stage of a written subtraction. Look at this one.

$$
\begin{array}{r}
{}^5\!\!\not{6} \quad {}^{9}\!\not{}{}^{\not{1}}\!\not{0} \quad {}^1 4 \quad 5 \\
-\quad 3 \quad\ 1 \quad\ 7 \quad\ 2 \\
\hline
2 \quad\ 8 \quad\ 7 \quad\ 3 \\
\hline
\end{array}
$$

Look carefully at what you must do if you want to borrow a number when the next column has a zero.

You have to borrow from the 1000s to ten 100s, then borrow one 100 to get ten 10s.

Here are some top tips for accurate subtraction work!

💡 Tips

- Check your subtractions by adding your answer to the number you subtracted.
- Try to estimate your answer before you start. It will help you to know if your answer is 'about right'.
- You can't have lots of subtractions in a list the way you can with additions. If you have to subtract two or three numbers, you must subtract one number and then subtract the next number from the answer.

💬 Talk maths

Look at this subtraction and explain it aloud. Say what is happening at each stage. Make sure you work in the correct order, right to left.

```
  ¹2̸   ¹¹2̸   ¹3    8
−   1     3     5    6
─────────────────────
          8     8    2
```

✔ Check

1. Complete each of these subtractions.

a.
	3	6	5
−	1	4	7

b.
	6	5	3
−	2	2	8

c.
	3	2	7
−	1	6	5

d.
	4	7	2	5
−	1	9	0	7

2. Complete each of these subtractions using a written method on squared paper.

a. 415 − 236   b. 824 − 375   c. 3542 − 937   d. 6042 − 3555

⚠ Problems

Brain-teaser

This chart shows the number of people attending three village fetes.

Village	Plink	Plank	Plonk
Number of visitors	825	936	89

a. How many more people went to Plink than Plonk? _____

b. How many more people went to Plank than Plink? _____

Brain-buster

This chart shows the number of fans at three rock concerts.

Band	Crush	Push	Mush
Fans	6455	7106	3453

If the fans of Push and Mush are combined, how many more people are there than fans of Crush? _____

Times tables facts

↻ Recap

You need to learn and to understand your tables.

Remember: to use the times tables grid, you find a number on the side and a number on the top, like 3 × 4, and where the row and the column meet you get your answer. So 3 × 4 = 12, easy!

×	1	2	3	4	5	6	7	8	9	10	11	12
1	1	2	3	4	5	6	7	8	9	10	11	12
2	2	4	6	8	10	12	14	16	18	20	22	24
3	3	6	9	12	15	18	21	24	27	30	33	36
4	4	8	12	16	20	24	28	32	36	40	44	48
5	5	10	15	20	25	30	35	40	45	50	55	60
6	6	12	18	24	30	36	42	48	54	60	66	72
7	7	14	21	28	35	42	49	56	63	70	77	84
8	8	16	24	32	40	48	56	64	72	80	88	96
9	9	18	27	36	45	54	63	72	81	90	99	108
10	10	20	30	40	50	60	70	80	90	100	110	120
11	11	22	33	44	55	66	77	88	99	110	121	132
12	12	24	36	48	60	72	84	96	108	120	132	144

You should already know your 2-, 3-, 4-, 5-, 8- and 10-times tables. Use the times tables grid to check your knowledge.

We can also check divisions on the grid, because division is the inverse of multiplication.

5 × 4 = 20, so 4 × 5 = 20. 20 ÷ 5 = 4, and 20 ÷ 4 = 5. Three numbers, four different facts.

📄 Revise

You should now be ready to learn your 6-, 7-, 9-, 11- and 12-times tables.

Times tables are like counting in steps. Can you see these on the grid above?

6-times table:	0	6	12	18	24	30	36	42	48	54	60	66	72
7-times table:	0	7	14	21	28	35	42	49	56	63	70	77	84
9-times table:	0	9	18	27	36	45	54	63	72	81	90	99	108
11-times table:	0	11	22	33	44	55	66	77	88	99	110	121	132
12-times table:	0	12	24	36	48	60	72	84	96	108	120	132	144

Remember that times tables are like counting in steps: 9 + 9 + 9 = 3 × 9 = 27

💡 Tips

- Often by looking at the 1s you can spot patterns. Look at the 9- and the 12-times tables above.
- And remember, multiplication works both ways, so if you know 6 × 7 you also know 7 × 6.

Some tables are easier to learn than others. Most people find the 2-, 5-, 10- and 11-times tables the easiest.

Talk maths

Play *In a minute*. You will need the times tables grid and a clock or watch.
Work with a partner. Take turns asking and answering times tables questions as fast as possible.

The questioner can use a hidden times tables grid to help them check answers quickly. How many questions can players answer in one minute?

If you prefer, you can set limits, such as the 6-times table only.

DID YOU KNOW?

The times tables grid only contains 78 different facts. (That's because many of the facts are repeated, such as 7 × 8 = 8 × 7). You already know thousands of facts, surely another 78 can't be that hard...

What is 4 × 6?

What is 63 ÷ 7?

What is 8 × 11?

What is 81 ÷ 9?

What is 96 ÷ 12?

✔ Check

1. Answer these times tables questions.

   a. 3 × 7 = _____   **b.** 5 × 9 = _____   **c.** 8 × 4 = _____

2. Answer these times tables questions.

   a. 20 ÷ 5 = _____   **b.** 24 ÷ 2 = _____   **c.** 42 ÷ 7 = _____

3. Write down all the numbers in the 6-times table.

   _____

4. Write down all the numbers in the 12-times table.

   _____

5. Which numbers are in both the 6- and 12-times tables?

   _____

⚠ Problems

Brain-teaser Josie is thinking of a number.
She says it is in the 6-times table and the 7-times table. What is the number? _____

Brain-buster Jim buys nine bags of crisps for 60p each. Explain how he can use his times tables to find the total cost of the crisps, and then give the answer.

_____

Mental methods for multiplication and division

↻ Recap

Multiplication is like repeated addition.
5 × 3 is 'five lots of 3' or 3 + 3 + 3 + 3 + 3.

Multiplication can be done in any order.
5 × 3 = 3 × 5 (answer = 15)

Division is equal sharing.
15 ÷ 3 is 15 shared into three equal lots.

Division is the inverse of multiplication.
2 × 9 = 18 so 18 ÷ 9 = 2 (and 18 ÷ 2 = 9)

DID YOU KNOW?
You cannot divide a number by zero.
The answer is infinity!

And multiplication is the *inverse* of division!

📄 Revise

Some calculations can be done mentally, without writing anything down.

You should know any calculation that is in the times tables.
3 × 5 = 15   4 × 9 = 36   15 ÷ 5 = 3   36 ÷ 4 = 9

And if you know these facts you can do other calculations.
30 × 5 = 150   4 × 90 = 360   150 ÷ 5 = 30   360 ÷ 4 = 90

Multiplying by 2 is just doubling; dividing by 2 is halving.
39 × 2 = 30 + 30 + 9 + 9 = 60 + 18 = 78   220 ÷ 2 = $\frac{1}{2}$ of 200 + $\frac{1}{2}$ of 20 = 100 + 10 = 110

When you multiply or divide by 10 or 100, all the digits move one or two places in the place value table.
23 × 10 = 230   4 × 100 = 400   320 ÷ 10 = 32   600 ÷ 100 = 6

When you multiply three numbers, they can be done in any order. It's easiest to multiply by the smallest number last.
2 × 6 × 8 = 2 × 48 = 96

Don't forget: anything multiplied by 1 remains the same, and anything multiplied by zero will equal zero.

💡 Tips

If you are feeling confident, you can mix and match the mental methods shown above. Look at these.

- 50 × 30 is the same as 5 × 10 × 3 × 10.
  Swap the order = 5 × 3 × 10 × 10 = 15 × 100 = 1500.
- 6200 ÷ 20 is done in two stages. 6200 ÷ 10 = 620, and then divide by 2 to give 620 ÷ 2 = 310.

Talk maths

Look at the different mental methods shown on the page opposite. Using a pencil and paper, secretly write down a few multiplications and divisions using the methods, or use the ones in this box.

Read each calculation aloud to a partner and challenge them solve it. If they make errors, explain where they went wrong.

6×8 $72 \div 9$ 150×2
$800 \div 100$ $320 \div 10$
12×200
20×30 $4 \times 5 \times 6$
$60 \div 30$ $10 \times 20 \times 30$
$160 \div 20$ $120 \div 40$

✔ Check

1. Solve these multiplications mentally.

   a. $36 \times 2 =$ _____
   b. $2 \times 2 \times 6 =$ _____
   c. $95 \times 10 =$ _____
   d. $3 \times 100 =$ _____
   e. $36 \times 20 =$ _____
   f. $3 \times 400 =$ _____
   g. $2 \times 4 \times 300 =$ _____
   h. $50 \times 90 \times 2 =$ _____

2. Solve these divisions mentally.

   a. $26 \div 2 =$ _____
   b. $64 \div 8 =$ _____
   c. $150 \div 10 =$ _____
   d. $3000 \div 100 =$ _____
   e. $120 \div 10 =$ _____
   f. $60 \div 20 =$ _____
   g. $300 \div 15 =$ _____
   h. $640 \div 80 =$ _____

⚠ Problems

Brain-teaser There are 120 children in a school.

a. If they each spend 20p at the tuck shop, how much money is collected altogether? _____

b. If the children are divided equally into 10 groups, how many will be in each group? _____

Brain-buster There are 120 children in a school.

a. If each child spends 20p at the tuck shop every day for three days, how much money will be collected altogether? _____

b. If the children are divided equally into 30 groups, how many will be in each group? _____

Written methods for short multiplication

↻ Recap

We know that multiplication is repeated addition.
$4 \times 7 = 7 + 7 + 7 + 7 = 28$

We have also looked at some mental methods for doing multiplications in your head.
$12 \times 20 = 12 \times 10 \times 2 = 120 \times 2 = 240$

Sometimes calculations are just too hard to do in your head.
1247×6

That's when it's time to use a **formal written method**.

📋 Revise

We call this short multiplication.

We know that our number system uses these place-value columns: 1000s, 100s, 10s and 1s.

To make multiplication easier, we can use formal written methods using these columns.

When multiplying a larger number by a number less than 10, you must multiply each digit at the top by the single digit. If necessary, exchange 1s, 10s or 100s, and then add them.

Notice how $4 \times 5 = 20$. We carry the two 10s and leave zero 1s in the 1s column. Also $4 \times 3 = 12$. We carry the one 100, then in the 10s column we add the two carried 10s to the two 10s and write 4 in this column. $4 \times 2 = 8$ but we need to add the carried 100 which makes 9, so we write 9 in the 100s column. So the answer is 940.

	2	3	5
×			4
	9	4	0
		1	2

💡 Tips

Make sure you know your times tables facts!

- To carry out short multiplication, you still need to know your multiplication facts.
- And don't forget, you can estimate your answer. Use this to check that your formal written method gives you the size of answer you expect.

  For example, 124×6 will be a more than 720 but less than 780.
  ($6 \times 120 = 720$, $6 \times 130 = 780$)

Talk maths

Remember: you still have to multiply zeros, and anything times zero is... zero!

Look at the short multiplications below and explain them aloud. Say how each stage was done.

a.

```
      3   4
×         4
─────────────
  1   3   6
      1
```

b.

```
      4   6   2
×             6
─────────────────
  2   7   7   2
      3   1
```

✔ Check

1. Complete each of these short multiplications.

a.
```
    3  8
×      3
────────
```

b.
```
    4  5  7
×         2
───────────
```

c.
```
    3  4  2
×         4
───────────
```

d.
```
    8  0  6
×         5
───────────
```

2. Complete these calculations using a written method for short multiplication.

a. 153 × 3   **b.** 267 × 2   **c.** 538 × 4   **d.** 1273 × 6

⚠ Problems

Brain-teaser A school's tuck shop sells muesli bars for 7p each.

143 bars are sold. How much money is collected? _____

Brain-buster The school tuck shop also sells cartons of juice for 20p each.

125 children buy a muesli bar and a carton of juice.
How much money is collected altogether? _____

Written methods for short division

↻ Recap

Dividing means sharing something equally. If we share six biscuits between three people, they will get two biscuits each.

We know that division is the **inverse** of multiplication, so your times tables give you lots of division facts.

$3 \times 5 = 15$, so $15 \div 3 = 5$ and $15 \div 5 = 3$

$6 \times 8 = 48$, so $48 \div 6 = 8$ and $48 \div 8 = 6$

$10 \times 12 = 120$, so $120 \div 12 = 10$ and $120 \div 10 = 12$

> Remember what 'divide' means. It tells you how many times one number 'goes into' another number.

📋 Revise

Sometimes we have to divide larger numbers, such as $348 \div 3$. If this is too difficult to do mentally, we can use a written method. Look at how we write this down.

$$\begin{array}{r} 1\ 1\ 6 \\ 3\overline{\smash{)}3\ 4\ ^18} \end{array}$$

- To begin, we divide 3 into the 3. This goes once so we write 1 above the 3.
- Then we look at the 4. 3 goes into 4 once with one remaining. So we put 1 above the 4 and carry the remaining 1 to the 8 to make 18.
- 3 goes into 18 six times, so we write 6 above 18.
- So, 348 divided by 3 equals 116.

> Check it with a multiplication! $3 \times 116 = 348$

💡 Tips

- Sometimes you might not be able to divide into the first number, so you need to carry it to the next number.

$$\begin{array}{r} 0\ 4\ 2 \\ 3\overline{\smash{)}1\ ^12\ 6} \end{array}$$

💬 Talk maths

Look at the short division below and explain it aloud, saying how each stage was done.

```
      1   3   5
5 | 6  ¹7  ²5
```

✔ Check

1. Complete each of these short divisions.

   a. 5 | 1 2 5

   b. 3 | 7 2

   c. 5 | 9 0

   d. 4 | 5 3 2

   e. 2 | 2 3 3 4

   f. 7 | 7 4 9

2. Use a written method for short division.

   a. 116 ÷ 2   b. 215 ÷ 5   c. 426 ÷ 3
   d. 616 ÷ 4   e. 372 ÷ 6   f. 927 ÷ 9

⚠ Problems

Brain-teaser There are 628 children in Piqwarts School. They are divided equally into four different House Teams.

How many children are there in each House? _____

Brain-buster Jen and her mum are going on a Swimathon. They have to swim 2760 metres in 5 hours. How many metres will they swim each hour if they swim at a steady pace?

_____ metres each hour

Equivalent fractions

↻ Recap

A fraction of a whole is an amount less than 1. It shows the whole divided into **equal** parts.

Two halves make a whole.

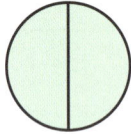

Four quarters make a whole.

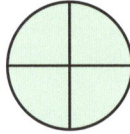

Three thirds make a whole.

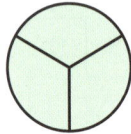

Six sixths make a whole.

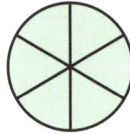

> Two quarters is equivalent to one half. We say that $\frac{2}{4}$ and $\frac{1}{2}$ are equivalent fractions. They represent the same amount.

> Two sixths is equivalent to one third. We say that $\frac{2}{6}$ and $\frac{1}{3}$ are equivalent fractions. They represent the same amount.

📋 Revise

We can use diagrams to show equivalent fractions, such as fraction walls.

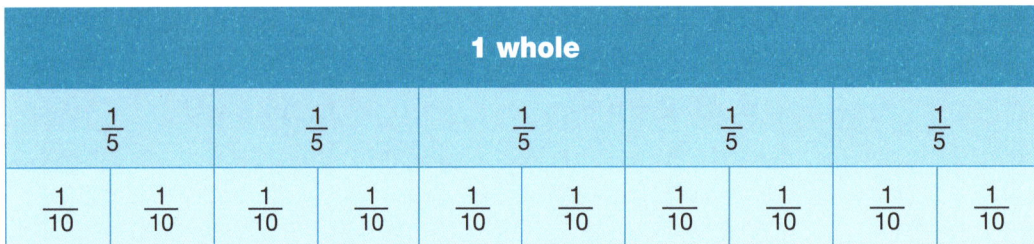

1 whole									
$\frac{1}{5}$		$\frac{1}{5}$		$\frac{1}{5}$		$\frac{1}{5}$		$\frac{1}{5}$	
$\frac{1}{10}$	$\frac{1}{10}$	$\frac{1}{10}$	$\frac{1}{10}$	$\frac{1}{10}$	$\frac{1}{10}$	$\frac{1}{10}$	$\frac{1}{10}$	$\frac{1}{10}$	$\frac{1}{10}$

Five fifths make a whole.
Ten tenths make a whole.

The wall shows us that two tenths is equivalent to one fifth, four tenths are equivalent to two fifths, and so on.

💡 Tips

> This makes it much easier to spot equivalent fractions!

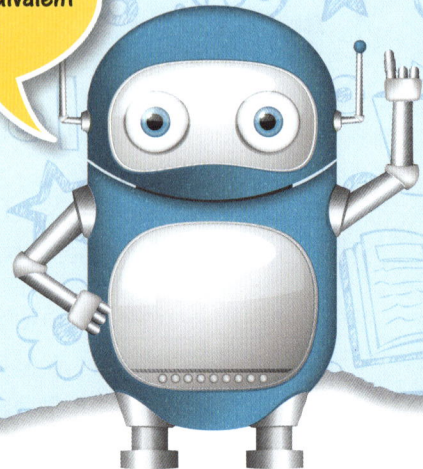

- Remember: when you divide a shape into fractions, every part must be the same size.

💬 Talk maths

Working with a partner, use counters or paper to cover up fractions in the fraction wall. Then say what fraction has been covered and what its equivalents are.

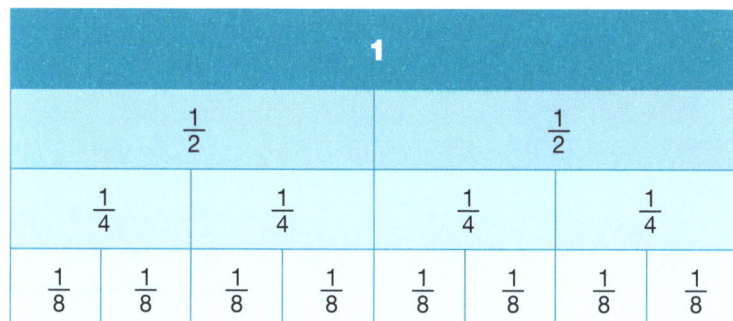

1							
$\frac{1}{2}$				$\frac{1}{2}$			
$\frac{1}{4}$		$\frac{1}{4}$		$\frac{1}{4}$		$\frac{1}{4}$	
$\frac{1}{8}$	$\frac{1}{8}$	$\frac{1}{8}$	$\frac{1}{8}$	$\frac{1}{8}$	$\frac{1}{8}$	$\frac{1}{8}$	$\frac{1}{8}$

Try making your own fraction wall, with one whole, thirds, sixths and ninths.

> Two quarters is equivalent to one half.

> Six eighths is equivalent to three quarters.

✔ Check

1. Shade each circle to show the fraction.

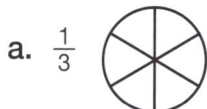

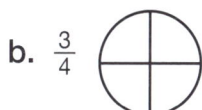

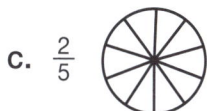

   a. $\frac{1}{3}$

   b. $\frac{3}{4}$

   c. $\frac{2}{5}$

   d. $\frac{3}{4}$

2. Draw a line to join each fraction in the top row to its equivalent below.

$\frac{1}{2}$	$\frac{1}{3}$	$\frac{1}{4}$	$\frac{1}{6}$

$\frac{3}{12}$	$\frac{2}{12}$	$\frac{4}{12}$	$\frac{6}{12}$

3. Circle the correct equivalent fraction.

   a. $\frac{3}{4}$: $\quad \frac{5}{8} \quad \frac{6}{8} \quad \frac{7}{8} \quad \frac{8}{8}$

   b. $\frac{2}{3}$: $\quad \frac{8}{9} \quad \frac{7}{9} \quad \frac{6}{9} \quad \frac{5}{9}$

   c. $\frac{5}{8}$: $\quad \frac{8}{16} \quad \frac{9}{16} \quad \frac{10}{16} \quad \frac{11}{16}$

   d. $\frac{3}{5}$: $\quad \frac{8}{20} \quad \frac{10}{20} \quad \frac{12}{20} \quad \frac{14}{20}$

⚠ Problems

Brain-teaser Jane is racing against Paul. Jane has finished $\frac{3}{4}$ of the race and Paul has finished $\frac{5}{8}$.

Who is nearer to the finish?

_____

Brain-buster Tina is racing against Joe. Tina has finished $\frac{7}{10}$ of the race and Joe has finished $\frac{4}{5}$.

Who is nearer to the finish?

_____

Adding and subtracting fractions

↺ Recap

| $\frac{1}{10}$ | $\frac{1}{10}$ | $\frac{1}{10}$ | $\frac{1}{10}$ | $\frac{1}{10}$ | $\frac{1}{10}$ | $\frac{1}{10}$ | $\frac{1}{10}$ | $\frac{1}{10}$ | $\frac{1}{10}$ |

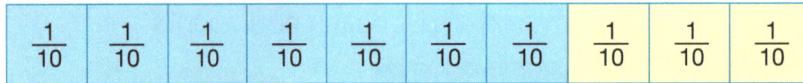

$$\frac{7}{10} + \frac{3}{10} = \frac{10}{10}$$

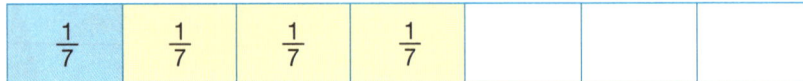

| $\frac{1}{7}$ | $\frac{1}{7}$ | $\frac{1}{7}$ | $\frac{1}{7}$ | | | |

$$\frac{1}{7} + \frac{3}{7} = \frac{4}{7}$$

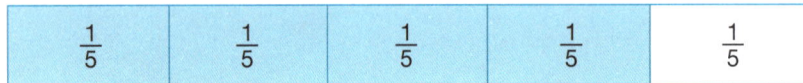

| $\frac{1}{5}$ | $\frac{1}{5}$ | $\frac{1}{5}$ | $\frac{1}{5}$ | $\frac{1}{5}$ |

$$\frac{5}{5} - \frac{4}{5} = \frac{1}{5}$$

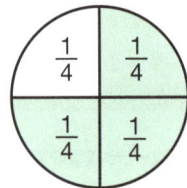

$$\frac{3}{4} - \frac{1}{4} = \frac{2}{4}$$

> You can add and subtract fractions that have the same denominator. Just add the numerator.

📄 Revise

Improper fractions have a numerator bigger than their denominator.

We can still add and subtract improper fractions, but you must still only add and subtract the numerators.

$\frac{5}{2} + \frac{6}{2} = \frac{11}{2}$ Five halves plus six halves equals eleven halves.

$\frac{3}{4} + \frac{5}{4} = \frac{8}{4}$ Three quarters plus five quarters equals eight quarters.

$\frac{7}{3} - \frac{5}{3} = \frac{2}{3}$ Seven thirds minus five thirds equals two thirds.

$\frac{9}{6} - \frac{4}{6} = \frac{5}{6}$ Nine sixths minus four sixths equals five sixths.

💡 Tips

- You can add three or more fractions, just like you can add three or more whole numbers. $\frac{2}{3} + \frac{4}{3} + \frac{5}{3} + \frac{3}{3} = \frac{14}{3}$

Talk maths

You can practise subtraction too: 'one whole minus three quarters'.

Remember:

A whole always has the same denominator and numerator.

$\frac{7}{7} = 1$ whole $\frac{3}{3} = 1$ whole $\frac{625}{625} = 1$ whole

Challenge a partner to find the rest of the whole.

For example, if you say 'six eighths', they must reply 'plus two eighths makes a whole'. If you say 'seven twelfths', your partner must say 'plus five twelfths makes a whole'.

Try it with larger fractions. 'Nine twentieths' ... 'plus eleven twentieths makes a whole' and so on.

✔ Check

1. What do you add to each of these fractions to make one whole?

   a. $\frac{1}{2} +$ _____ = 1   b. $\frac{3}{4} +$ _____ = 1   c. $\frac{1}{3} +$ _____ = 1   d. $\frac{3}{7} +$ _____ = 1

2. Complete these subtractions from one whole.

   a. $1 - \frac{1}{2} =$ _____   b. $1 - \frac{2}{5} =$ _____   c. $1 - \frac{7}{8} =$ _____   d. $1 - \frac{13}{20} =$ _____

3. Add these fractions.

   a. $\frac{3}{4} + \frac{2}{4} =$ _____   b. $\frac{4}{5} + \frac{3}{5} =$ _____   c. $\frac{6}{10} + \frac{3}{10} =$ _____

   d. $\frac{5}{8} + \frac{6}{8} =$ _____   e. $\frac{2}{7} + \frac{4}{7} + \frac{5}{7} =$ _____   f. $\frac{5}{6} + \frac{4}{6} + \frac{3}{6} =$ _____

4. Subtract these fractions.

   a. $\frac{5}{3} - \frac{1}{3} =$ _____   b. $\frac{5}{6} - \frac{2}{6} =$ _____   c. $\frac{13}{8} - \frac{7}{8} =$ _____

   d. $\frac{7}{4} - \frac{1}{4} =$ _____   e. $\frac{4}{5} - \frac{1}{5} =$ _____   f. $\frac{21}{20} - \frac{8}{20} =$ _____

⚠ Problems

Brain-teaser A pizza is cut into 12 equal slices. Tina eats half of it, Josie eats $\frac{2}{12}$ and Dan eats $\frac{3}{12}$.

What fraction of the pizza is left? _____

Brain-buster Some children share a bucket of popcorn. Tom takes $\frac{3}{20}$ and Amanda takes $\frac{6}{20}$.

What fraction of the popcorn is left? _____

Tenths and hundredths

↺ Recap

There are ten tenths in a whole.

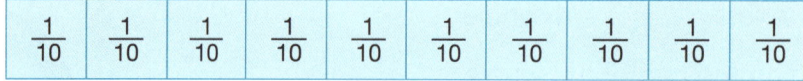

And there are 100 hundredths in a whole!

$\frac{1}{10}$	$\frac{1}{10}$	$\frac{1}{10}$	$\frac{1}{10}$	$\frac{1}{10}$	$\frac{1}{10}$	$\frac{1}{10}$	$\frac{1}{10}$	$\frac{1}{10}$	$\frac{1}{10}$

📋 Revise

We get tenths when we divide an object or number by 10.

One pizza is divided between 10 people.
Each person will receive $\frac{1}{10}$ of a pizza.

Now imagine if $\frac{1}{10}$ of the pizza was shared between 10 other people.

$\frac{1}{10}$ divided by 10 will give each person $\frac{1}{100}$ of the pizza. Not much!

We can count in tenths and hundredths.

Ten hundredths equal one tenth.

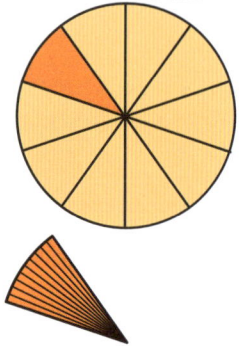

What would you rather have: one tenth of a pizza or one hundredth of a pizza?

					$\frac{1}{10}$				
$\frac{1}{100}$	$\frac{1}{100}$	$\frac{1}{100}$	$\frac{1}{100}$	$\frac{1}{100}$	$\frac{1}{100}$	$\frac{1}{100}$	$\frac{1}{100}$	$\frac{1}{100}$	$\frac{1}{100}$

$\frac{10}{100} = \frac{1}{10}$ $\frac{20}{100} = \frac{2}{10}$ $\frac{30}{100} = \frac{3}{10}$ $\frac{40}{100} = \frac{4}{10}$ $\frac{50}{100} = \frac{5}{10}$

$\frac{60}{100} = \frac{6}{10}$ $\frac{70}{100} = \frac{7}{10}$ $\frac{80}{100} = \frac{8}{10}$ $\frac{90}{100} = \frac{9}{10}$ $\frac{100}{100} = \frac{10}{10}$

💡 Tips

- Remember $\frac{50}{100}$ is equivalent to $\frac{5}{10}$, and both are the same amount as $\frac{1}{2}$.
- $1 \div 10 = \frac{1}{10} =$ one tenth

  $1 \div 100 = \frac{1}{100} =$ one hundredth

💬 Talk maths

Play *Beat the clock*.

Working with a partner, take turns to say any fraction in tenths or hundredths, such as two tenths or seventeen hundredths. Your partner then has to count on 10 more tenths or hundredths in less than 30 seconds.

Seven tenths.

Eight tenths, nine tenths, ten tenths, eleven tenths...

✔ Check

1. Add these tenths and hundredths.

   a. $\frac{1}{10} + \frac{6}{10} =$ _____

   b. $\frac{7}{10} + \frac{5}{10} =$ _____

   c. $\frac{13}{10} + \frac{9}{10} =$ _____

   d. $\frac{31}{100} + \frac{7}{100} =$ _____

   e. $\frac{78}{100} + \frac{3}{100} =$ _____

   f. $\frac{120}{100} + \frac{60}{100} =$ _____

2. Subtract these tenths and hundredths.

   a. $\frac{9}{10} - \frac{4}{10} =$ _____

   b. $\frac{7}{10} - \frac{5}{10} =$ _____

   c. $\frac{23}{10} - \frac{9}{10} =$ _____

   d. $\frac{21}{100} - \frac{11}{100} =$ _____

   e. $\frac{58}{100} - \frac{6}{100} =$ _____

   f. $\frac{125}{100} - \frac{80}{100} =$ _____

3. Write these fractions in words.

   a. $\frac{6}{10}$ _____

   b. $\frac{9}{10}$ _____

   c. $\frac{14}{100}$ _____

   d. $\frac{91}{100}$ _____

4. Write these as fractions.

   a. seven tenths _____

   b. thirteen tenths _____

   c. thirty-five hundredths _____

   d. two hundredths _____

⚠ Problems

Brain-teaser Complete this sentence.

Three tenths = ☐ hundredths

Brain-buster Complete this sentence.

Sixty-three hundredths = ☐ tenths and ☐ hundredths

Fraction and decimal equivalents

↻ Recap

A fraction is a proportion of one whole.

$\frac{1}{100}$ $\frac{1}{10}$ $\frac{1}{4}$ $\frac{1}{2}$ $\frac{3}{4}$ are all fractions.

📄 Revise

100s	10s	1s	0.1s	0.01s
		.		

Numbers less than one can also be represented by decimals.

To show tenths and hundredths using our number system, we use a decimal point and two new columns.

- We write three tenths as 0.3 and five hundredths as 0.05.
- We can say that the number 0.47 has four tenths and seven hundredths.
- We read decimals aloud, using digits zero to nine.
- Any fraction can be written as a decimal.

Fraction	$\frac{1}{2}$	$\frac{1}{4}$	$\frac{3}{4}$	$\frac{1}{10}$	$\frac{4}{10}$	$\frac{61}{100}$	$\frac{73}{100}$
Decimal	0.5	0.25	0.75	0.1	0.4	0.61	0.73

We can also have whole numbers and decimals.

For 23.62 we would say twenty-three, six tenths and two hundredths.

We say 0.5 is 'zero point five'.

We say 0.75 is 'zero point seven five'.

Or twenty-three point six two.

💡 Tips

- Decimals with **one decimal place** are equivalent to a fraction with a denominator of 10.

  $0.6 = \frac{6}{10}$

- Decimals with **two decimal places** are equivalent to a fraction with a denominator of 100.

  $0.37 = \frac{37}{100}$

  0.1 is one tenth $(\frac{1}{10})$

  0.2 is two tenths $(\frac{2}{10})$

  0.3 is three tenths $(\frac{3}{10})$

Any decimal can be written as a fraction!

Can you keep going?

Talk maths

Practise saying these decimals both ways.

'Three tenths' or 'zero point three'.

0.3	0.17	0.4	0.08	0.85
0.75	0.2	0.02	0.31	0.11
0.1				
0.05	0.43	3.14	0.66	0.99

'One tenth and seven hundredths' or 'zero point one seven'.

✔ Check

1. Write the shaded part of each whole as a decimal.

   a.   b.   c.

2. Change these fractions to their decimal equivalents.

   a. $\frac{1}{2}$ = _____   b. $\frac{3}{4}$ = _____   c. $\frac{1}{10}$ = _____

   d. $\frac{27}{100}$ = _____   e. $\frac{1}{4}$ = _____   f. $\frac{8}{10}$ = _____

3. Change these decimals to their fraction equivalents.

   a. 0.25 = _____   b. 0.78 = _____   c. 0.4 = _____

   d. 0.75 = _____   e. 0.5 = _____   f. 0.21 = _____

⚠ Problems

Brain-teaser Tim says that zero point two is the same as $\frac{2}{10}$.

Explain why he is right. _____

Brain-buster Zoe says that seventy-five hundredths is the same as three quarters.

Explain if she is wrong or right. _____

103

Working with decimals

↺ Recap

Our number system uses **place value**.

We sometimes call this 100s, 10s and 1s.
346 is three hundred and forty-six.
Between zero and one we use decimal fractions.
.12 is point one two.

100s	10s	1s	0.1s	0.01s
3	4	6 •	1	2

Decimals show tenths and hundredths of a whole.
These are sometimes called decimal fractions.

DID YOU KNOW?

When you solve money problems with pounds and pence, you often use decimals.

📋 Revise

There are lots of ways we can work with decimals.

Dividing whole numbers by 10 or 100

We move the place value one column to the right when dividing by 10.
$\frac{3}{10} = 3 \div 10 = 0.3$

We move the place value two columns to the right when dividing by 100.
$\frac{45}{100} = 45 \div 100 = 0.45$

Look at these examples: $\frac{7}{10} = 0.7$ and $\frac{7}{100} = 0.07$

$\frac{23}{10} = 2.3$ and $\frac{23}{100} = 0.23$

Rounding decimals just like other numbers

Rounding to the nearest whole number: $0.7 \rightarrow 1$ $4.2 \rightarrow 4$ $6.8 \rightarrow 7$ $3.5 \rightarrow 4$

If it is .5 or higher, round up. If it is lower than .5, round down.

Ordering and comparing decimals

0.8 is bigger than 0.5 (eight tenths is bigger than 5 tenths).

0.25 is smaller than 0.32 (thirty-two hundredths is bigger than twenty-five hundredths).

💡 Tips

We can draw number lines for decimals too!

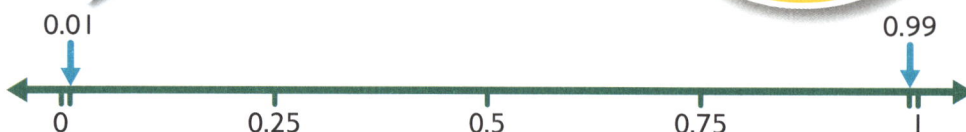

0.01 ↓   0.99 ↓

|————|————————|————————|————————|————|
0   0.25   0.5   0.75   1

Talk maths

Choose ten numbers between 10 and 100, like the ones in the box below.
Take any two of the numbers and divide each one by 100.
Next, make a true statement about your numbers.

| 12 | 19 | 23 | 37 | 42 | 69 |
| 53 | 84 | 95 | 75 | | |

For example: 23 and 75
$23 \div 100 = 0.23$ $75 \div 100 = 0.75$

0.23 is less than 0.75. 0.75 is greater than 0.23.

You can extend this activity by writing each decimal on a number line from 0 to 1.

✔ Check

1. Complete these divisions.

   a. $7 \div 10 =$ _____   **b.** $31 \div 100 =$ _____   **c.** $3 \div 10 =$ _____   **d.** $94 \div 100 =$ _____

2. Round these decimals to the nearest whole number.

   a. $4.8 =$ _____   **b.** $3.1 =$ _____   **c.** $5.5 =$ _____   **d.** $7.4 =$ _____

3. Insert the correct signs. Use <, > or =.

   a. 0.7 _____ 0.75   **b.** 0.31 _____ 0.42   **c.** 0.6 _____ 0.60   **d.** 0.25 _____ 0.23

4. Position these decimals on the number line.

   0.5   0.9   0.05   0.75   0.65   0.25   0.35

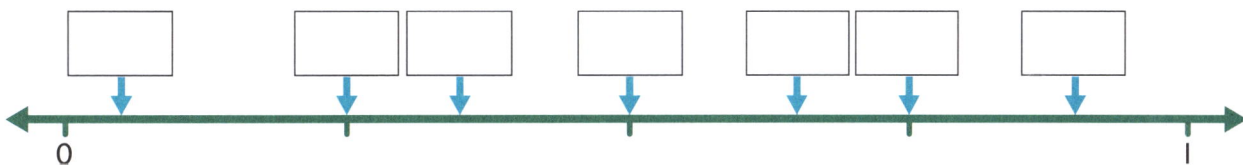

   0 _____ 1

⚠ Problems

Brain-teaser Brenda knows that 65 divided by 100 = 0.65.

What would 65 divided by 10 be? _____

Brain-buster Thomas says that 0.55 rounded to the nearest whole number is 1.

What would 0.55 rounded to the nearest tenth be? _____

Units of measurement

↺ Recap

Different quantities are measured in different ways.

Quantity	Units of measurement	Abbreviations	
Time (years)	1 year = 12 months 1 week = 7 days	years = y months = m	weeks = w days = d
Time (days)	1 day = 24 hours 1 hour = 60 minutes 1 minute = 60 seconds	hours = h minutes = m	seconds = s
Money	1 pound = 100 pence	pounds = £	pence = p
Mass	1 kilogram = 1000 grams	kilograms = kg	grams = g
Capacity	1 litre = 1000 millilitres	litre = l	millilitre = ml
Length	1 kilometre = 1000 metres 1 metre = 100 centimetres 1 centimetre = 10 millimetres	kilometres = km centimetres = cm millimetres = mm	metres = m

🗒 Revise

Think about where you see the different units above, and what they are used for.

Quantity	How to measure	Why	Examples
Time (years)	calendars, clocks	planning years, days	holidays, timetables
Money	notes and coins	buying and selling	school meals
Mass	scales and weights	to know correct amounts	making cakes
Capacity	containers	to know correct amounts	mixing drinks
Length	rulers, tape measures, maps	making things, planning journeys	making a box, going on a trip

💡 Tips

- How far do you walk to school?
- How many days have you been alive?
- Do you have a water bottle? What is its capacity?
- Guess the weight of your dinner. How can you check?
- How much pocket money do you receive each year?

Talk maths

Play *Read my mind*.
Take turns with a partner to give clues about the units you are thinking about. Try thinking of different clues for all the units on the opposite page.

> Can you think of more challenging clues?

Example clue	Answer
a cup of tea	millilitres
a long journey	kilometres
a finger	centimetres
a snail	grams
a new car	pounds
a maths lesson	minutes
a cricket match	hours
a lifetime	years

✔ Check

1. **What units would you measure these things in?**

   a. The length of a pencil _____
   b. The height of a house _____
   c. The cost of a pencil _____
   d. The cost of a car _____
   e. The capacity of a cup _____
   f. The capacity of a bath _____
   g. The weight of a computer _____
   h. The weight of a pencil _____
   i. The duration of pop song _____
   j. The duration of a holiday _____

2. **Write the names of this measuring equipment.**

   a.
   b.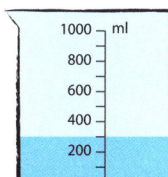
   c.

   _____   _____   _____

3. **What equipment would you use to measure these items?**

   a. The length of a pencil _____
   b. The height of your teacher _____
   c. The capacity of a cup _____
   d. The weight of an apple _____
   e. A running race _____
   f. The time until Christmas _____

⚠ Problems

Brain-teaser 1 millilitre of water weighs 1 gram.
How much does 5 litres of water weigh, in grams? _____

Brain-buster Tyler's dad is one metre seventy-three centimetres tall.
Tyler is exactly half the height of his dad. How tall is Tyler in millimetres? _____

Units of time

↻ Recap

These are all units of time: seconds, minutes, hours, days, weeks and years.

60 seconds = 1 minute
60 minutes = 1 hour
24 hours = 1 day
7 days = 1 week
365 days = 1 year

Remember that months are not all the same length.
30 days hath September, April, June and November.
All the rest have 31, except for February all alone.
(February has 28 days, and 29 days in a leap year.)

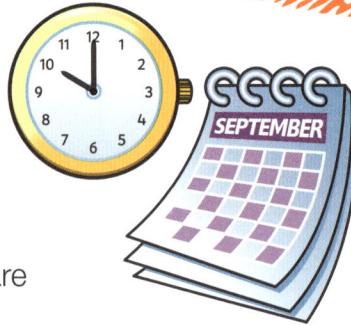

DID YOU KNOW?

The Earth actually takes 365 and a quarter days to travel around the sun. That's why we have a leap year every four years to catch up with the extra quarters.

Also, we say that there are 52 weeks in a year, but this is not exactly true, it is an approximation. Look: 7 × 52 = 364 days

📄 Revise

To convert between times is tricky!

Convert	Calculation	Example
years to months	× 12	3 years = 3 × 12 = 36 months
years to days	× 365	2 years = 2 × 365 = 730 days
weeks to days	× 7	6 weeks = 6 × 7 = 42 days
days to hours	× 24	3 days = 3 × 24 = 72 hours
hours to minutes	× 60	5 hours = 5 × 60 = 300 minutes
minutes to seconds	× 60	10 minutes = 10 × 60 = 600 seconds

💡 Tips

- 1 year = 365 days × 24 hours × 60 minutes × 60 seconds!

If you want to convert from a large unit, such as years, to a small unit such as seconds, you have to do one stage at a time.

💬 Talk maths

Work with an adult and think about how long different things take to do in seconds.

Task	Time	In seconds
putting on shoes	2 minutes	120 seconds
walking to school	$5\frac{1}{2}$ minutes	330 seconds

✔ Check

1. Draw lines to match the time in the top row to its equivalent on the bottom row.

2 years   2 weeks   2 days   2 hours   2 minutes

120 minutes   730 days   120 seconds   48 hours   14 days

2. Convert each of these times.

   a. 2 minutes = _____ seconds

   b. 3 hours = _____ minutes

   c. 4 days = _____ hours

   d. 5 weeks = _____ days

   e. 6 years = _____ months

   f. 2 years (not leap years) = _____ days

3. Convert each of these times.

   a. $3\frac{1}{2}$ minutes = _____ seconds

   b. $2\frac{1}{2}$ hours = _____ minutes

   c. $5\frac{1}{2}$ days = _____ hours

   d. $7\frac{1}{2}$ years = _____ months

⚠ Problems

Brain-teaser Poppy knows that on her tenth birthday she will have lived for three leap years and seven ordinary years.

How many days has she lived altogether? _____

Brain-buster a. How many hours are there in a non-leap year? _____

         b. How many minutes are there in a day? _____

Analogue and digital clocks

↺ Recap

When we use the 12-hour clock, we divide the day into two halves of 12 hours each, from midnight to noon, and then back to midnight.

For 12-hour clock time, we have to say am or pm.

24-hour clocks are different – they do just what they say. They start at midnight and count 24 hours through the day.

This is an analogue clock.
It uses hands to show the time.
It is a 12-hour clock.
It shows twenty-three minutes past ten, but is it am or pm?

am and pm are 'before noon' and 'after noon' to you and me!

You have to look out the window and see if it is day or night!

📋 Revise

Digital clocks use digits to show hours and minutes.

For 12-hour digital times, we have to use am and pm. This shows that the time is before noon or after noon.

Converting between 12-hour and 24-hour digital times isn't so hard.

For pm times, we add 12 to get the 24-hour time for example, 10:15pm = 22:15pm.

For 24-hour times past 12 noon, just subtract 12 to get the 12-hour time, for example 16:47 = 4:47pm.

11:35pm = 23:35
6:42pm = 18:42

13:25 = 1:25pm
22:45 = 10:45pm

Learn how to convert between analogue and digital.

💡 Tips

- 'Past' times are easy. We just write the hours and the number of minutes. For example, 10:05 is five minutes past 10.
- 'To' times are harder. Learn that 30 is half past, 40 is twenty to, 45 is quarter to and 50 is ten to.

Talk maths

Work with a partner to become an expert time-teller.

Draw an analogue clock like this. Put hands on the clock, using a long pencil and a short pencil, and say am or pm. Then challenge your partner to say this in 12-hour or 24-hour digital time.

Then write digital clock times, such as 17:35. Ask your partner to show you the time on the analogue clock.

✔ Check

1. Complete this chart for 12-hour analogue and digital times.

Analogue	twelve noon		ten past eleven	five to four	
Digital		8:45			3:15

2. Write these analogue times as 24-hour digital times.

a. am

b. pm

c. am

d. pm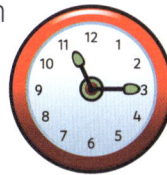

_____   _____   _____   _____

3. Write these 24-hour digital times as 12-hour analogue times.

a.

b.

c.

d.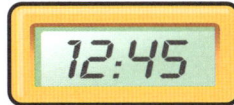

_____   _____   _____   _____

⚠ Problems

Brain-teaser A train departs at quarter to eleven in the morning and arrives at 12:05pm.

How long does the journey take? _____

Brain-buster An aeroplane takes off from London at 21:45 and flies directly to South Africa. The aeroplane lands at 9:25am London time.

How long was the flight? _____

Money

↻ Recap

These are the coins we use in England and Wales.
We also use £5, £10, £20 and £50 notes.

📄 Revise

Money shows us the cost of things.
We use pounds and pence.
£1 = 100 pence

We show pence using two decimal places.

7 pounds and 25 pence = £7.25

That's *seven pounds twenty-five*.

Unlike other decimals, if the last digit is a zero, we still write it in.

16 pounds and 50 pence = £16.50
That's *sixteen pounds fifty*.

Look at this amount: £0.59 is zero pounds and fifty-nine pence, or 59p

To convert pounds to pence, multiply by 100:
£6.50 = 6.50 × 100 = 650p

To convert pence to pounds divide by 100:
3265p = 3265 ÷ 100 = £32.65

DID YOU KNOW?

Before the year 1971 we used pounds, shillings and pence. A shilling was worth 12 old pennies and there were 240 old pennies in a pound!

1p is one hundredth of one pound.

Notice that if you use the £ sign and decimals, you don't add a p at the end.

💡 Tips

Operation	Example
Addition	£3.50 + £2.15 = £5.65
Subtraction	£5.00 − £1.25 = £3.75
Multiplication	£2.10 × 3 = £6.30
Division	£7.00 ÷ 2 = £3.50
Fractions	$\frac{1}{2}$ of £25.00 = £12.50

You can use all your number skills to solve money problems.

- We can use written methods with money just like any other numbers.
- Remember to be careful with the decimal point.

Talk maths

I'd like a tin of beans for 32p. Here is £1.

Here is your change: 68p.

Find an old shopping receipt or a price list from a catalogue or website, and work with a partner to compare costs.

Next, challenge each other by asking for items on the list and paying for them.

If you feel confident, put the money away and solve problems just using the maths.

✔ Check

1. Convert these pence to pounds.

Pence	500p	150p	3300p	59p	1000p
Pounds					

2. Convert these pounds to pence.

Pounds	£1	£4.25	£0.62	£20	£12.06
Pence					

3. Complete these calculations.

a. £2.50 + £3.30 = _____

b. £4.90 + £3.20 = _____

c. £10.00 − £6.50 = _____

d. £20.00 − £12.99 = _____

e. 8 × 50p = £ _____

f. £2.50 × 4 = £ _____

g. £20 ÷ 4 = £ _____

h. £15 ÷ 3 = £ _____

⚠ Problems

Brain-teaser Ice creams cost £1.25 each. Alfie's mum buys five.

a. What is the total cost? _____

b. How much change will she get from a £10 note? _____

Brain-buster Ice creams cost £1.25 each and ice lollies cost £1.50 each.

a. How much would three ice creams and seven lollies cost altogether? _____

b. How much change would there be from a £20 note? _____

Mass and capacity

↺ Recap

Mass tells us the weight of things.

We measure mass in grams and kilograms.
1000g = 1kg

Capacity tells us how much a container can hold.

We measure capacity in millilitres and litres.
1000ml = 1l

📄 Revise

To add masses, each mass must be in kilograms, or in grams.

The units must be the same.

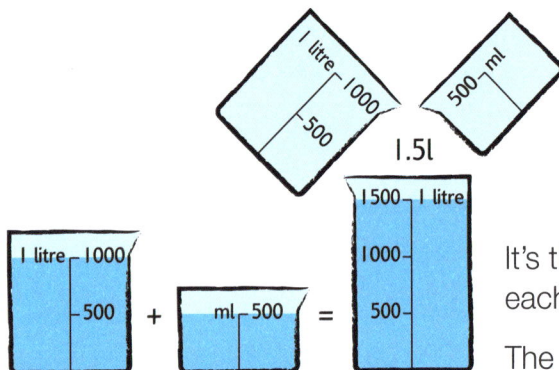

It's the same with capacity: to add or subtract capacities, each capacity must be in litres, or in millilitres.

The units must be the same.

💡 Tips

These charts may be useful for converting units.

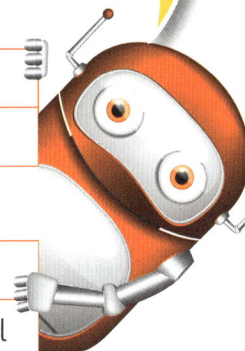

Converting mass

Kilograms	1kg	5kg	5kg 350g	$\frac{1}{4}$kg	$\frac{1}{2}$kg	$\frac{3}{4}$kg
Grams	1000g	5000g	5350g	250g	500g	750g

Converting capacity

Litres	1l	3l	4l 825ml	$\frac{1}{4}$l	$\frac{1}{2}$l	$\frac{3}{4}$l
Millilitres	1000ml	3000ml	4825ml	250ml	500ml	750ml

Talk maths

Working with an adult, look in the kitchen cupboards or visit a shop. Challenge each other to predict the capacity of different bottles, the mass of different bags of vegetables, and so on.

If you can, extend this work using scales and measuring jugs. Try to investigate the mass or capacity of as many different items as you can.

Remember to work safely and with adult permission at all times!

✔ Check

1. Match each capacity to the correct object.

thimble	mug	bathtub
200l	200ml	20ml

2. Match each mass to the correct object.

mouse	child	elephant
5000kg	50g	50kg

3. Convert these masses.

   a. 5kg = ____ g

   b. 6000g = ____ kg

   c. $\frac{1}{2}$ kg = ____ g

   d. 4500g = ____ kg

4. Convert these capacities.

   a. 3000ml = ____ l

   b. $7\frac{1}{2}$ l = ____ ml

   c. 3500ml = ____ l

   d. $\frac{1}{2}$ l = ____ ml

5. Solve these calculations.

   a. $3\frac{1}{4}$ kg + 500g = _____ kg

   b. 450g + 700g = _____ g

   c. $\frac{1}{10}$ kg + 320g = _____ g

   d. $3\frac{1}{4}$ l + 450ml = _____ ml

   e. 230ml + 140ml = _____ l

   f. $3\frac{1}{4}$ l + $1\frac{3}{4}$ l = _____ ml

⚠ Problems

Brain-teaser

a. A carton of juice contains 50ml. How many cartons would make 1 litre? _____

b. A pencil weighs 27g. How much would 100 pencils weigh?

   _____ kilograms and _____ grams

Brain-buster

a. Paper cups have a capacity of 100ml. How many cups will a $2\frac{1}{2}$ l bottle fill? _____

b. Onions weigh 125g each. How many onions would make 1 kilogram? _____

Length and distance

↻ Recap

We measure short lengths in centimetres and millimetres, and some longer lengths and distances in metres.

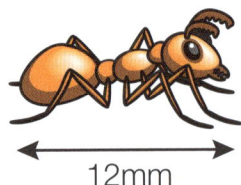

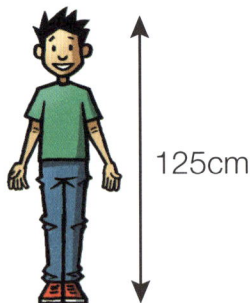

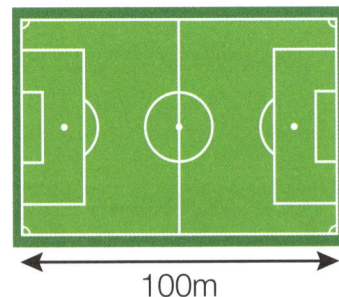

12mm

125cm

100m

📄 Revise

We measure even longer distances in kilometres.

You should know these. 10mm = 1cm

100cm = 1m (and 1000mm = 1m)

1000m = 1km

10km

A

B

Length is the measure of an object or a line from end to end.
For example, a football pitch is 100m long; a finger is 1cm wide.

Distance is the measure of the space between two points or two objects.
For example, the distance between two towns is 10km; the gap between two parked cars is 1m.

Remember tenths and hundredths?
A millimetre is one tenth of a centimetre, a centimetre is one hundredth of a metre.

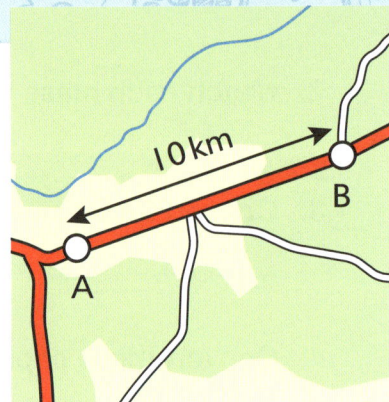

💡 Tips

These charts may be useful to help you convert units. Remember, when you add lengths together, they must have the same units!

Millimetres	1	5	10
Centimetres	0.1	0.5	1

Centimetres	1	10	25	50	75	100
Metres	0.01	0.1	0.25	0.5	0.75	1

Metres	10	100	250	500	750	1000
Kilometres	$\frac{1}{100}$	$\frac{1}{10}$	$\frac{1}{4}$	$\frac{1}{2}$	$\frac{3}{4}$	1

💬 Talk maths

Working with a partner, challenge each other, without using a ruler, to draw a line of a particular length, or make two dots a certain distance apart.

(Never go above 12cm, but try mixtures of centimetres and millimetres, such as 5.7cm.)

Next measure the length or distance and see how accurate your partner is.

✔ check

1. Measure the length of these lines. Give your answers in millimetres.

a. _____ _____   b. _____ _____

c. _____ _____   d. _____ _____

2. Measure the distances between these dots. Give your answers in centimetres.

a. A •   • B _____

b. A •   • B _____

c. A •   • B _____

3. Complete these conversion charts.

a.
mm	cm
10	
100	
	2
	35
	100

b.
cm	m
100	
1000	
	0.25
	0.5
	10

c.
m	km
500	
2000	
	$\frac{1}{4}$
	1
	9

⚠ Problems

Brain-teaser Ahmed is 1.52m tall. He wears shoes with a 2cm heel.

What height will he be with his shoes on? _____ m or _____ cm

Brain-buster The distance between two towns is 19km. Isobel is going to ride her bike between the towns, and wants to stop halfway. How far will she ride for each half of the journey?

Give your answer in km and in m: _____ km or _____ m

117

Perimeter

↻ Recap

Perimeter is the distance around the sides of a shape.

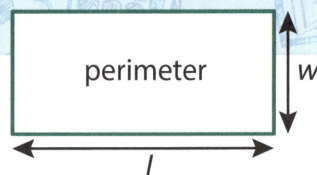

perimeter w l

Don't forget to always show the units!

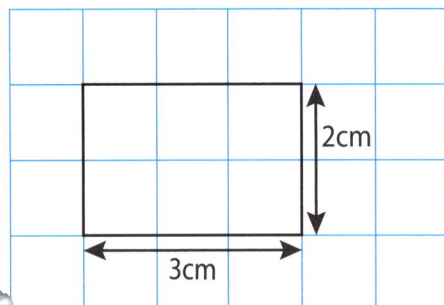

2cm
3cm

4cm
4cm

This rectangle has a perimeter of 3 + 3 + 2 + 2 = 10cm.

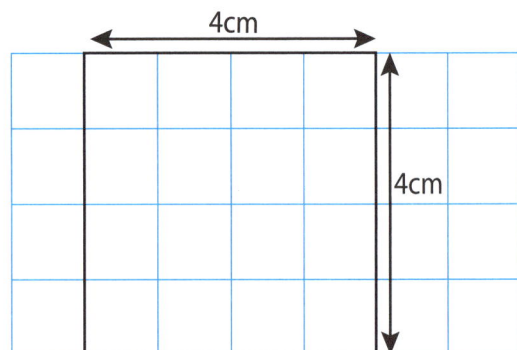

This square has a perimeter of 4 + 4 + 4 + 4 = 16cm.

目 Revise

If we say that all rectangles have a length l and a width w, then the perimeter can be calculated with a formula.

We can say $P = l + w + l + w$

Or, changing the order that we add in: $P = l + l + w + w$

We can make this simpler too: $P = 2l + 2w$

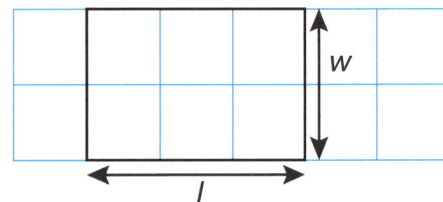

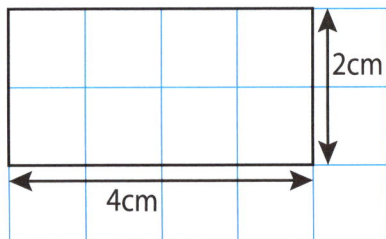

w
l

2cm
4cm

The perimeter of this rectangle is
$P = 2 \times 4 + 2 \times 2 = 12$cm

What do you think the perimeter of a rectangle 6m long and 3m wide would be?

💡 Tips

Here's a quick way to find the perimeter of a square.

- The formula for a square is easier, because all the sides are the same length.

  $P = 4s$

- $P = 4 \times 3 = 12$cm

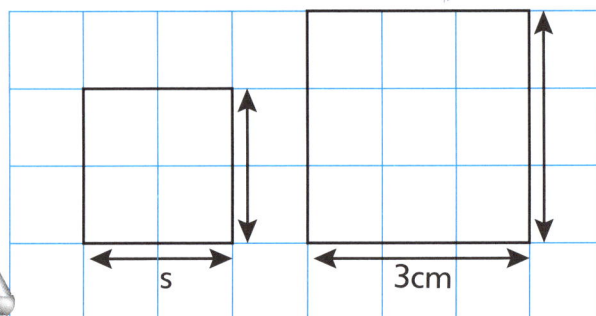

s
3cm

💬 Talk maths

Use a pencil and a ruler to draw a selection of rectangles and squares, each with different measurements. Label your shapes A, B, C and write the lengths of the sides separately.

Next, challenge someone you know to estimate the perimeter of each shape. Use a ruler to check their estimates. How many can they get right?

✔ Check

1. **Draw each of these shapes and then write their perimeter inside them.**

   a. A rectangle, length 4cm, width 2cm.

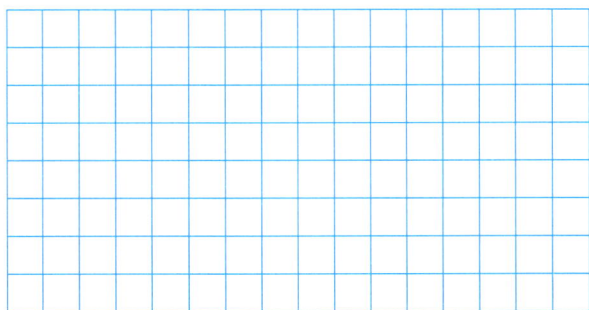

   b. A square, side length 3cm.

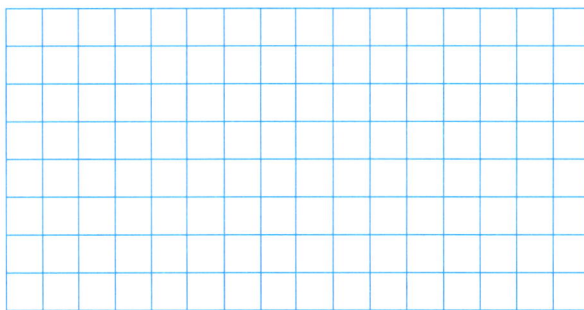

2. **Calculate the perimeter of these shapes.**

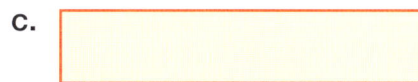

   a.

   b.

   c.

   _____   _____   _____

3. **Complete this chart.**

Shape	Length	Width	Perimeter
rectangle	5cm	2cm	
rectangle	12mm	5mm	
rectangle	6km	2km	
square	8mm	8mm	
square	5m	5m	
square	1.5cm	1.5cm	

⚠ Problems

Brain-teaser A square field is $3\frac{1}{2}$ km long on each side.

How long is the fence that goes around it? _____

Area

↻ Recap

A 2D shape is flat. 2D means two-dimensional.

We can accurately draw 2D shapes on paper with a pencil and a ruler.

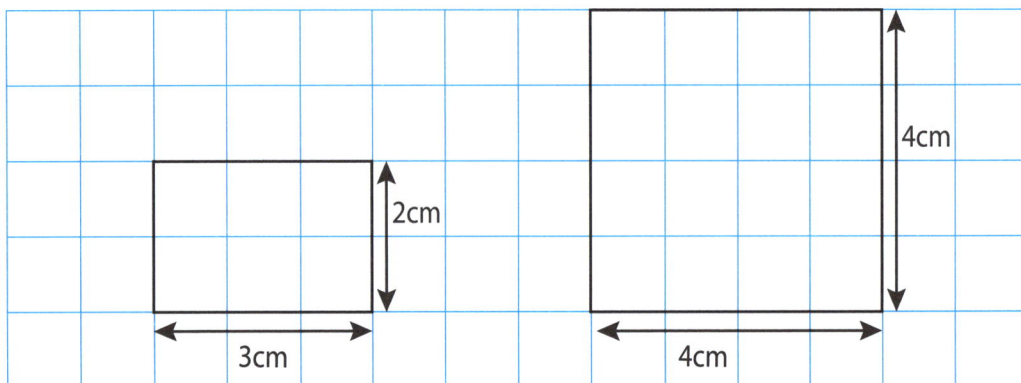

2cm

3cm

4cm

4cm

Remember, a rectangle has right angles at each corner, and the opposite sides are the same length.

A square is a type of rectangle, where all sides are the same length.

📄 Revise

Area is measured in squares.

This square is 1 square long and 1 square high. We say its area is 1 square.

We can simply count squares to calculate simple areas.

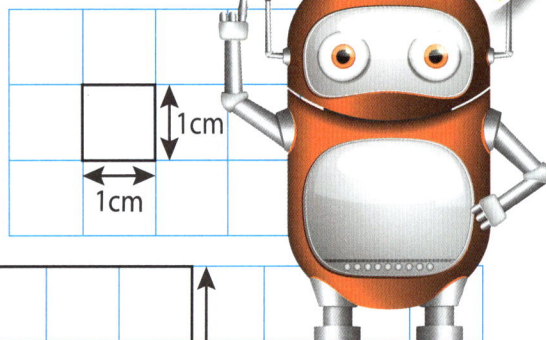

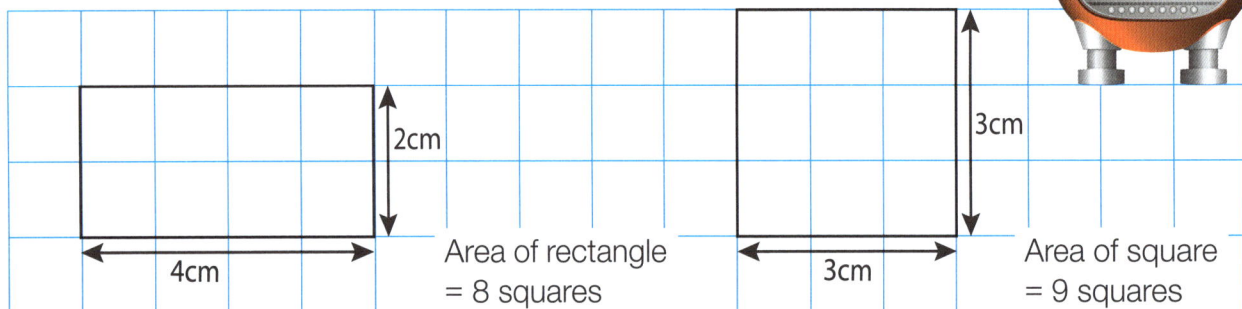

1cm

1cm

2cm

4cm

Area of rectangle = 8 squares

3cm

3cm

Area of square = 9 squares

💡 Tips

- This shape is 2 squares wide and 3 squares long. It looks like an array that we use in multiplication. We could write 2 × 3 for the array. We can use this multiplication fact to work out the area.

There are 2 lots of 3 squares. **So the area is 6 squares.**

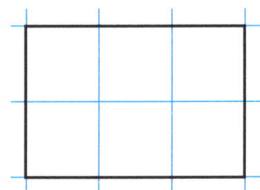

💬 Talk maths

Some of these have more than one answer. Can you find them all?

Work with a partner to draw each of these shapes, if possible on squared paper.

Before you start discuss what you think the length and height of each shape will be.

Shape	rectangle	rectangle	rectangle	square	square	square
Area	6 squares	8 squares	12 squares	1 square	4 squares	16 squares
Length						
Width						

✔ Check

1. Count squares to find the area of each shape.

a.

b.

c.

d.

2. Find the area of these shapes. You can draw them on squared paper to help you.

a. A rectangle that is 4 squares long and 3 squares high: _____

b. A square that is 3 squares long and 3 squares high: _____

⚠ Problems

Brain-teaser Sanjay draws a rectangle 12 squares long and 7 squares high, and a square with each side 8 squares long. Which shape has the greater area, and by how much?

_____

Brain-buster Mark makes a wall from wooden blocks. It is 3 blocks high and 4 blocks long. It has a gap that is 1 block wide and 2 blocks high.

How many blocks does Mark use for his wall? _____

Angles

↻ Recap

Angles are used to measure how much things turn.

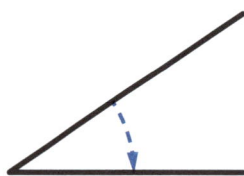

- A quarter of a turn is called a right angle.
- A half-turn is two right angles.
- Three quarters of a turn is three right angles.
- There are four right angles in a complete turn.

Angles are also used to measure the gap where two straight lines meet.

This angle is smaller than a right angle.

📄 Revise

Acute: less than a right angle

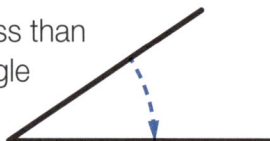

Right angle:

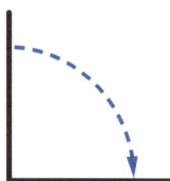

Obtuse: more than a right angle, less than two right angles

✔ Check

1. What is each angle: an acute angle, an obtuse angle, or a right angle?

a.

b.

c.

_____   _____   _____

2. Number these angles a, b, c and d, going from smallest to largest.

a.

b.

c.

d.

_____   _____   _____   _____

⚠ Problems

Brain-teaser Draw a triangle and cut it out. Cut off each angle.
Put them together so that the points are touching. What do you notice? _____

Triangles

↻ Recap

A triangle is a 2D shape with three straight sides. It also has three angles.

- Angles less than a quarter turn are called **acute**.
- A **right angle** is a quarter turn.
- Angles between one and two right angles are called **obtuse**.

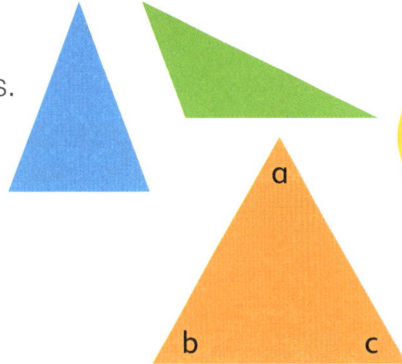

Tri means three, so triangle means three angles!

📄 Revise

There are several types of triangle.

Equilateral	Isosceles	Right-angled	Scalene
All sides are the same length. All angles are the same size.	Two sides are the same length. Two angles are the same.	One angle is a right angle.	All sides are different lengths. All angles are different sizes.

✔ Check

1. Name each of these triangles.

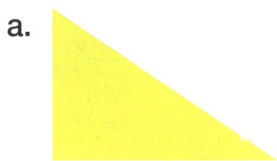

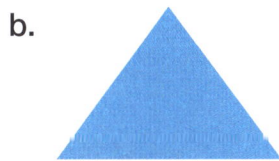

a.

b.

_____

_____

c.

d.

_____

_____

⚠ Problems

Brain-teaser Draw one straight line to make an isosceles triangle and a scalene triangle.

Quadrilaterals

↻ Recap

We say that different 2D shapes have different properties.

Triangle	Square	Rectangle	Pentagon	Hexagon
3 sides	4 sides	4 sides	5 sides	6 sides

▤ Revise

A square is a **regular** quadrilateral, and an equilateral triangle is a regular triangle. All their sides and angles are the same. **Irregular** shapes do not have equal sides or angles.

There are six types of quadrilateral.

Square	Rectangle	Rhombus	Parallelogram	Kite	Trapezium
All sides equal. All angles right angles.	Opposite sides equal. All angles right angles.	All sides equal. Opposite angles equal.	Opposite sides equal and parallel. Opposite angles equal	Adjacent sides equal.	Only one pair of parallel sides.

Adjacent means 'next to'.

💡 Tips

Can you see the connection between a square and a rhombus?

- Try making quadrilaterals with construction kits, straws or lolly sticks. You can see how they can be stretched and squashed to make other quadrilaterals.

💬 Talk maths

Cover the names on the opposite page and practise naming each quadrilateral. Then cover the shapes and try to describe the properties for each name.

✔ Check

1. What is the difference between a square and a rectangle?

_____

2. What is the difference between a rhombus and a kite?

_____

3. What is the difference between a parallelogram and a trapezium?

_____

4. Name each quadrilateral, and then connect it to its properties.

a. _____   b. _____   c. _____   d. _____

All sides equal. Opposite angles equal.

Adjacent sides equal.

Opposite sides equal and parallel. Opposite angles equal.

Only one pair of parallel sides.

⚠ Problems

Brain-teaser Petra drew a quadrilateral with all sides the same length.

Which quadrilaterals could she have drawn? _____

Brain-buster Roger wants to draw a kite and thinks he can do this by joining a right-angled triangle and an isosceles triangle together. Draw a sketch to show how he could do this.

Symmetry of 2D shapes

↻ Recap

Some objects have two identical halves.

When one half is like a mirror image of the other half, we can say the object or shape is symmetrical.

Revise

Look at the butterfly above. It has a **line of symmetry**. This is a bit like a mirror. One side is an identical reflection of the other.

Some objects and shapes have their lines of symmetry in different positions.

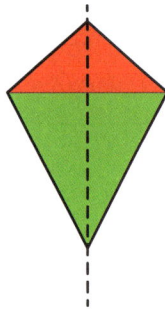

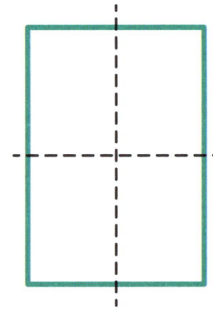

Lines of symmetry can be harder to spot. Can you see them for these two?

This diagram has two lines of symmetry.

This square has four lines of symmetry.

💡 Tips

- Remember that for patterns, colours must be symmetrical too.
- Regular shapes have the same number of lines of symmetry as they do sides.
- A circle has an infinite number of lines of symmetry.
- The letters Z and S might look symmetrical, but their halves are not mirror images.

Talk maths

With a partner, discuss these shapes and identify all the lines of symmetry. Many of them only have one line, but some have more than one.

✔ Check

1. Carefully complete these shapes to make them symmetrical.

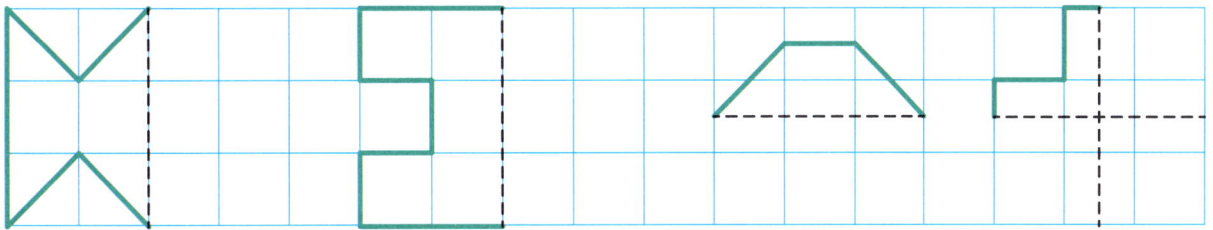

2. Draw the lines of symmetry on each number that has one or more.

1 2 3 4 5 6 7 8 9 0

3. Draw all the lines of symmetry on these regular shapes.

⚠ Problems

Brain-teaser Draw the other half of this shape to make it symmetrical.

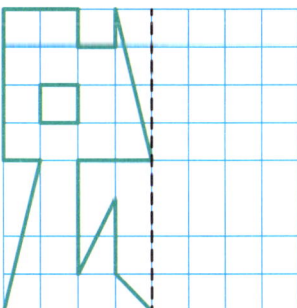

Brain-buster Tanya says she can think of four capital letters that each have two or more lines of symmetry. Can you show each letter with its lines of symmetry?

Coordinates

↻ Recap

Number lines are used for counting in equal amounts.

They can be drawn in any direction.

📋 Revise

We draw graphs with a vertical *y*-axis and a horizontal *x*-axis.
Each one is like a number line. They meet at zero.

We can plot points on the grid using coordinates. Points on the grid are always plotted with the *x*-coordinate first, and then the *y*-coordinate.

The coordinates of point G are (4, 5).

That's 4 along the x-axis, and 5 up the y-axis.

The coordinates of point H are (6, 3).

Can you find the coordinates for points J and K?

J = (_____ , _____), K = (_____ , _____)

We can join the points to form lines.
For the line AB, A = (3, 7), B = (8, 8)

We can also plot the corners of shapes.
Can you find the coordinates of the vertices of the triangle XYZ?

X = (_____ , _____), Y = (_____ , _____), Z = (_____ , _____)

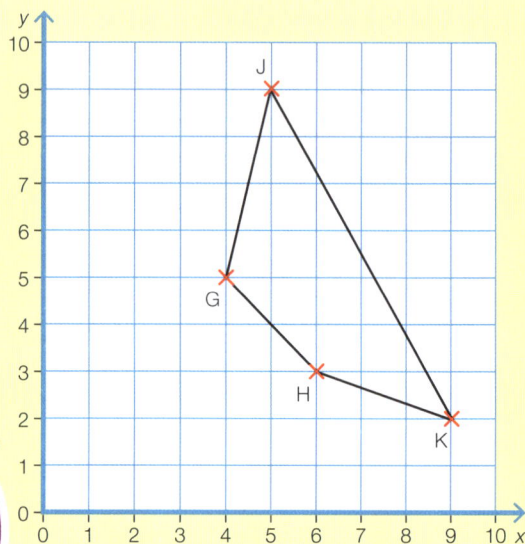

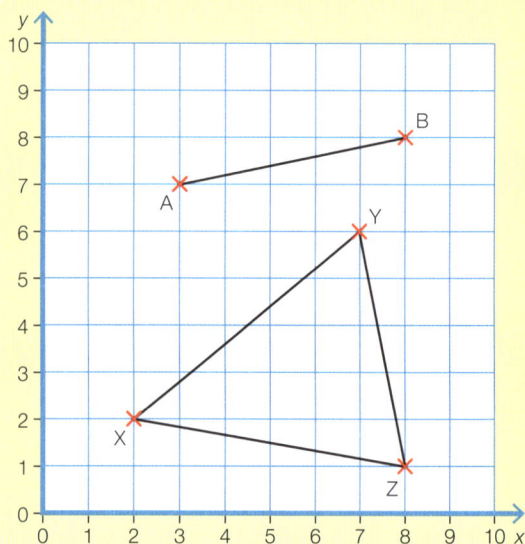

💡 Tips

- Remember, when you are plotting points, or reading and writing coordinates, along first and then up.
- Some people say 'along the corridor and up the stairs'.

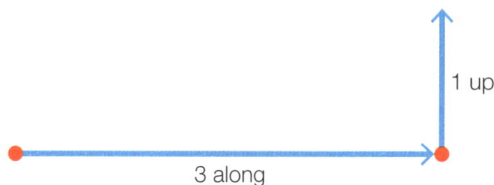

1 up

3 along

Talk maths

Challenge a partner to find the treasure at different coordinates.

Next, send them on a treasure hunt, sending them on a walk to five or six different points on the map.

The treasure is buried at (5, 4).

Start at (0, 0), next go to (6, 2), then to (3, 5)... Can your partner follow the directions?

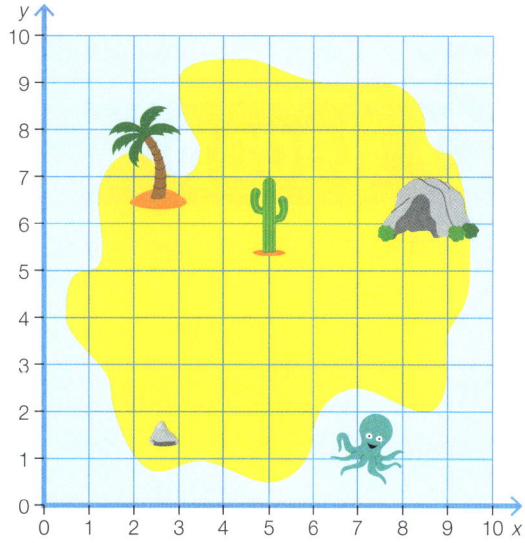

✔ Check

1. a. Use the grid below. Write the coordinates for these points.

X = (_____ , _____), Y = (_____ , _____)

b. Plot these points.

A (2, 2)   B (2, 8)   C (8, 8)   D (8, 2)

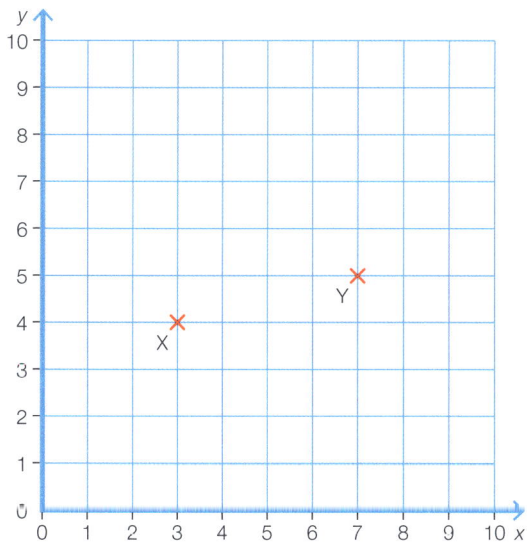

2. a. For the graph below. Write the coordinates of the triangle PQR.

P = (_____ , _____), Q = (_____ , _____),
R = (_____ , _____)

b. Draw a rectangle with vertices JKLM with these coordinates.

J (1, 1)   K (9, 1)   L (9, 7)   M (1, 7)

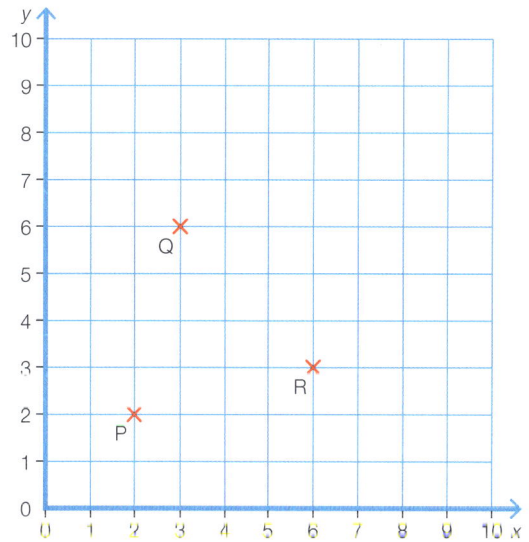

⚠ Problems

Brain-teaser Tim plots a point (7, 2). What are the coordinates of a point three squares directly above it? (_____ , _____)

Translation

↺ Recap

We draw graphs with a y-axis and an x-axis.
We can join the points to form lines.
We can also plot the corners of shapes.
We write the x-coordinate first, then the y-coordinate.
A = (1, 2), X = (2, 4)

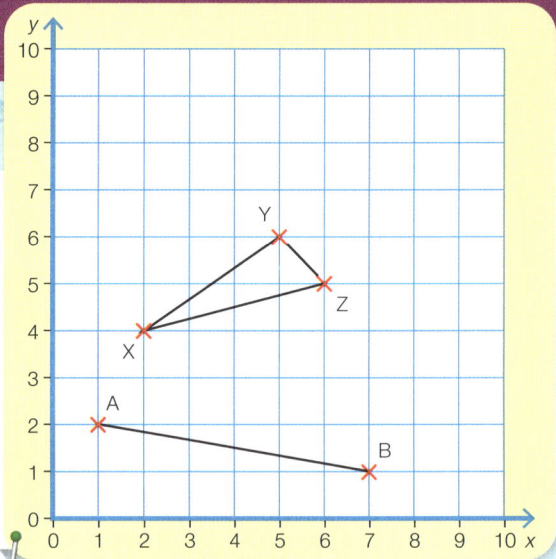

Remember: along, then up.

Translation is when all of the shape moves the same distance.

📋 Revise

We can **translate** points. We can move a point on the grid, with the x-coordinate and the y-coordinate each moving a certain amount.

The red point A below has been translated 3 right and 4 up. The new point has the coordinates (5, 6).

Can you see how the red points B, C and D have been translated?

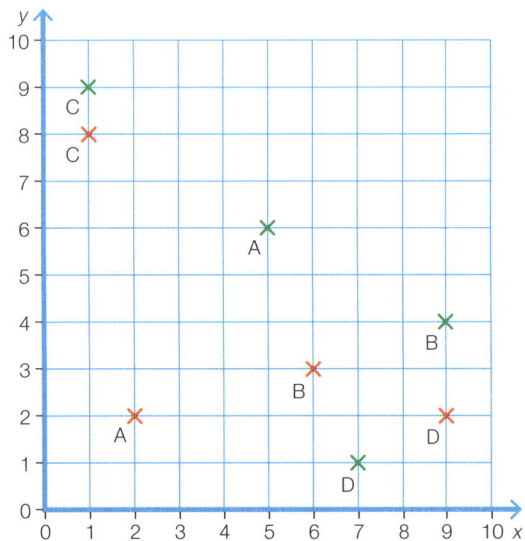

We can also translate shapes.

The square below has been translated 4 left and 3 down. Can you see how each corner has moved the same amount?

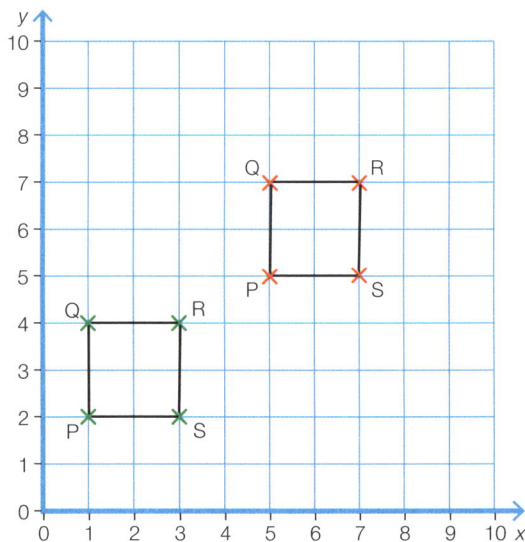

💡 Tips

- When translating a shape, all the x-coordinates change by the same amount, and all the y-coordinates do too.

Talk maths

Take turns to choose different points on a graph. Say its coordinates, and then challenge someone to translate it.

Translate (1, 1) by 2 right and 3 up.

Translate (8, 8) by 0 left and 2 down.

✔ Check

1. a. Using the coordinate grid below, translate the points W, X , Y and Z by 2 right and 3 down.

b. Write the coordinates of the new points.

W = (___ , ___)   X = (___ , ___)

Y = (___ , ___)   Z = (___ , ___)

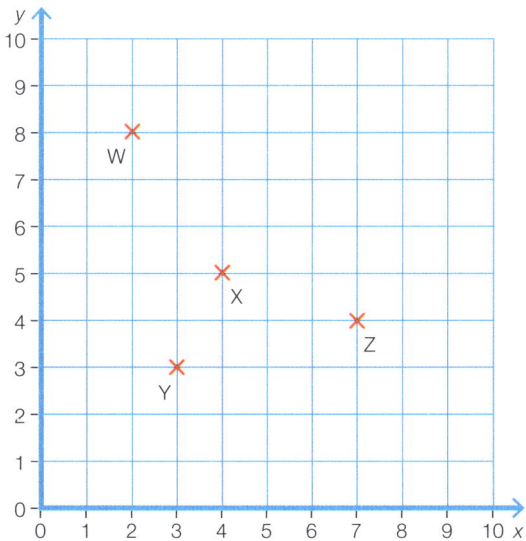

2. Use the coordinate grid below.

a. Plot a triangle ABC: A (2, 2), B (4, 4), C (4, 2).

b. Translate it 3 right and 5 up.

c. Write the coordinates of the new shape here.

A = (___ , ___)   B = (___ , ___)

C = (___ , ___)

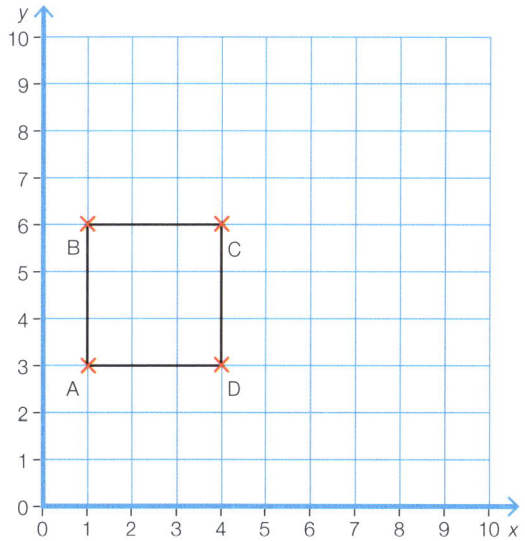

⚠ Problems

Brain-teaser Tina draws a grid and marks point X with coordinates (5, 4).

She then translates it to (0, 0). What was the translation? (_____ , _____)

Tables and pictograms

↻ Recap

Pictograms provide information in an easy-to-read way. They use simple icons to represent data.

Look at this pictogram. It was created after a class survey of pets. It uses one icon for each pet counted in the survey.

Cat	☺ ☺ ☺ ☺ ☺
Dog	☺ ☺ ☺ ☺ ☺ ☺
Goldfish	
Budgie	☺ ☺ ☺
Hamster	☺ ☺ ☺

☺ = 1

- The children in this class have 17 pets altogether.
- There is the same number of budgies as hamsters.
- There are no goldfish.
- There is one more dog than cats.

📋 Revise

Complicated information is often presented in tables and charts.
This table provides information about some animals in a safari park.

Animal	Height (cm)	Weight (kg)	Average lifespan (years)	Diet
Elephant	350	5000	58	herbivore
Giraffe	530	1200	24	herbivore
Lion	100	160	17	carnivore
Rhinoceros	150	1500	45	herbivore
Zebra	130	320	23	herbivore

Notice that each column has different units.

Looking down each column we can compare the information for different animals. For example, we can say about the lion:
The lion is the only carnivore. It is the shortest animal. It is the lightest animal.

We can also do calculations. For example: *The lion is 50cm shorter than the rhinoceros.*

💡 Tips

- Use your fingers to help you read rows and columns of data. Or if you have a ruler available even better as it is easy to misread tables and charts.

💬 Talk maths

Work with a friend to do a survey. You could do either of these.

- Create a table about your friends or families. Include information about their ages, heights, eye colour and so on.
- Create a pictogram of car colours.

When you have your data, discuss it, and ask each other questions about it.

✔ Check

Children's weather survey for a term.

Weather	Icon	Days
Sunshine	☀	10
Rain	💧	25
Cloudy	☁	30
Snow	❄	5

1. **Draw a pictogram for the weather survey below.**

   Use one icon for every 5 days.

Sunshine						
Rain						
Cloudy						
Snow						

2. **Use the animal information table on the opposite page to answer these questions.**

a. Which is the heaviest animal? _____

b. Which animals are shorter than a rhinoceros? _____

c. Which two animals have a similar lifespan? _____

d. How much taller is the giraffe than the elephant? _____

e. What is the difference in weight between the rhinoceros and the giraffe? _____

f. How much longer on average does a zebra live than a lion?

⚠ Problems

Brain-teaser Tom thinks of two animals that are listed in the table opposite. He says one is twice as heavy as the other.

Which two animals is he thinking about? _____

133

Bar charts

↻ Recap

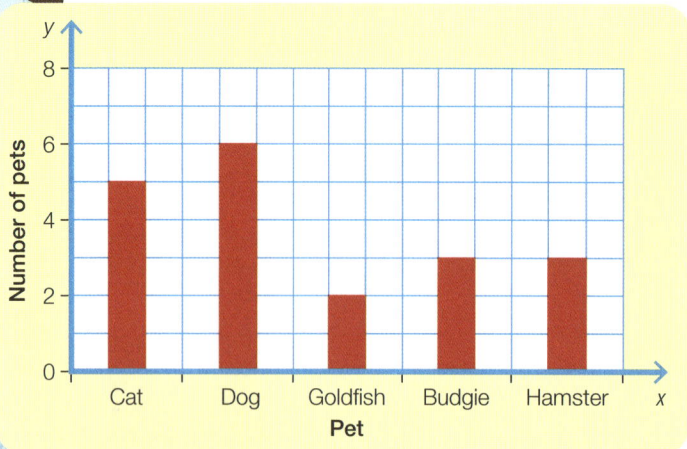

Bar charts are like pictograms in simpler form.

Instead of icons they use bars to represent the different quantities, with a scale on one axis to show the number of each item's bar.

This is a bar chart for pets in a class. Notice that the scale increases in 2s.

📋 Revise

This is a bar chart for a car survey outside a school. The vertical axis is in 5s.

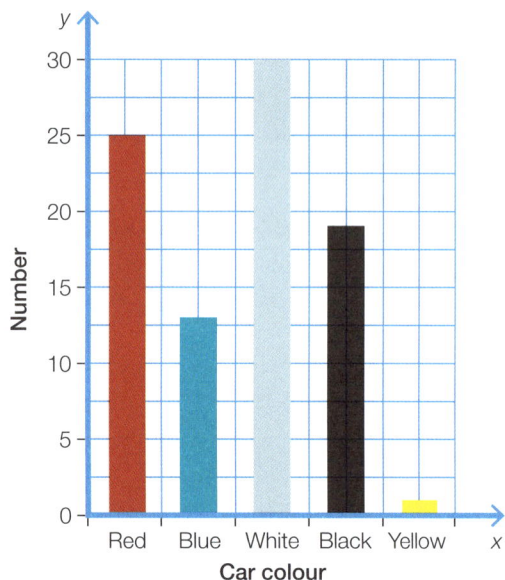

This is a bar chart for the number of pupils at four local schools. The vertical axis is in 100s.

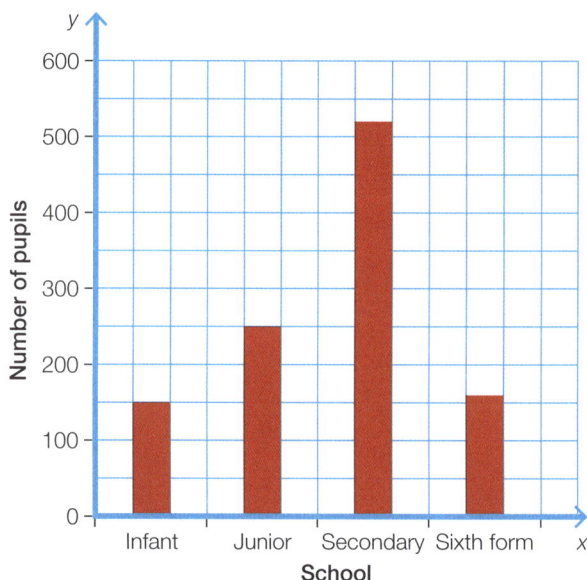

Look at the tables of data. Can you see how each number is shown on the bar chart?

Car colour	Red	Blue	White	Black	Yellow
Number	25	13	30	19	1

Now do the same for the numbers of pupils.

School	Infant	Junior	Secondary	Sixth form
Pupils	150	250	520	160

Talk maths

You could draw a bar chart for favourite colours.

Work with a partner to try to list as many different things that you could make bar charts for. Think of surveys you might do at school or sports that you like.

Discuss what scale you might need to use.

✔ Check

A group of friends count the number of books they have at home. Then they make a bar chart. Use the bar chart to answer these questions.

1. Who has the most books?

   _____

2. How many books does Karl have?

   _____

3. How many books are there altogether? _____

4. How many more books does Ella have than Ravi? _____

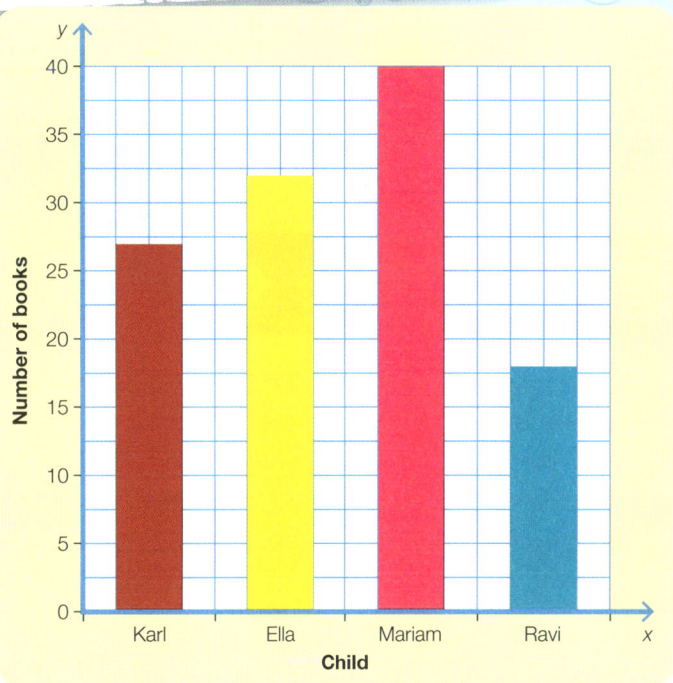

⚠ Problems

Brain-teaser Create a bar chart for the animal lifespan information below. Choose your scale carefully.

Animal	Average lifespan (years)
Elephant	58
Giraffe	24
Lion	17
Rhinoceros	45
Zebra	23

Time graphs

Recap

Each of these graphs has a vertical y-axis and a horizontal x-axis.

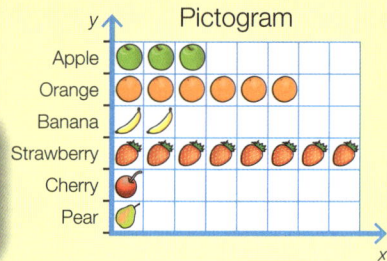

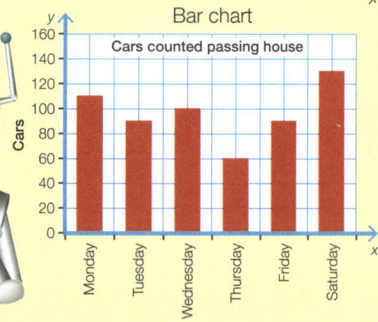

We can represent information and data in different types of charts and graphs.

Bar charts and pictograms are useful for presenting information from surveys.

- How do you travel to school?
- What is your favourite snack?
- Do you have any pets?

Pictogram

Bar chart

Cars counted passing house

Revise

Time graphs are useful for showing how things change over time, such as temperature changing or things growing.

This graph shows how the temperature changed in the playground during a school day. Look at how you can draw lines to find the temperature at any time of day.

Read the graph to check the table below. Can you read the temperature for 4pm?

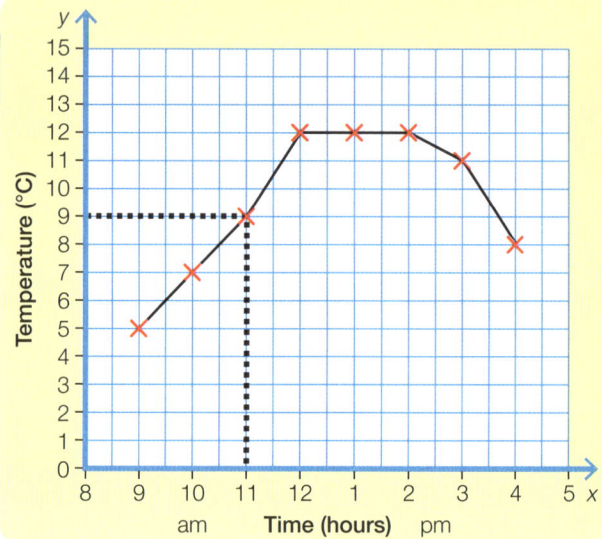

Time	9am	10am	11am	12noon	1pm	2pm	3pm
Temperature	5°C	7°C	9°C	12°C	12°C	12°C	11°C

Tips

- You can create line graphs by plotting points and then joining them.
- This chart shows the height of a tree each year for 5 years.

Time (years)	1	2	3	4	5
Height (metres)	1	3	4	5	5

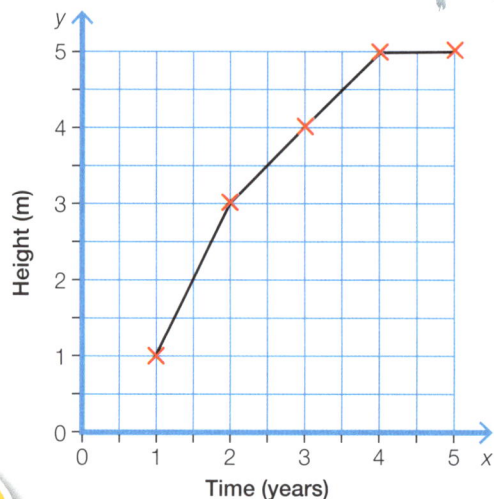

Remember how to plot coordinates: along, and then up.

💬 Talk maths

Working with an adult, use this graph to draw a time graph of your own. (It is OK to invent data rather than find real data.)

- What will the units of time be: seconds, minutes, hours, days, weeks or years?
- What will you show changing in time: a person's height, a baby's weight, the depth of water in a bath, the distance of a bike ride?
- What units will you use for the quantity that is changing: kilograms, litres, kilometres?

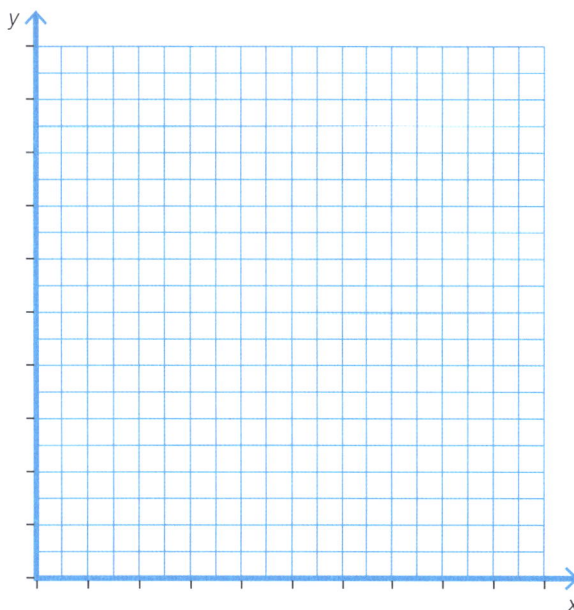

✔ Check

Use the graph opposite to answer these questions.

1. At what time was the water coldest?

   _____

2. When was the water 4°C?

   _____

3. Find the difference between the warmest and coldest temperatures. _____

Sea temperature at night

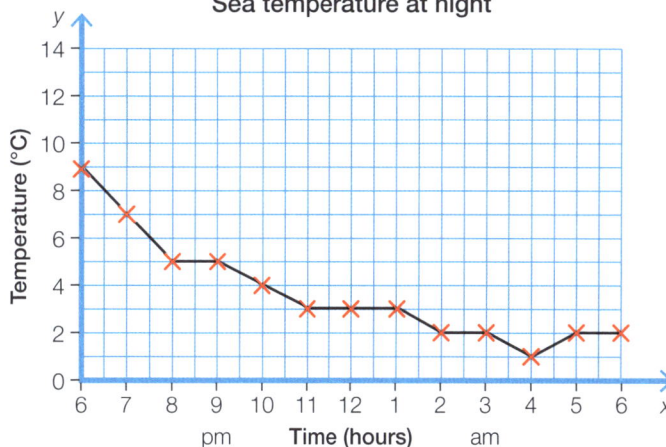

⚠ Problems

Month	0	1	2	3	4	5	6
Height (cm)	37	40	42	46	47	48	50

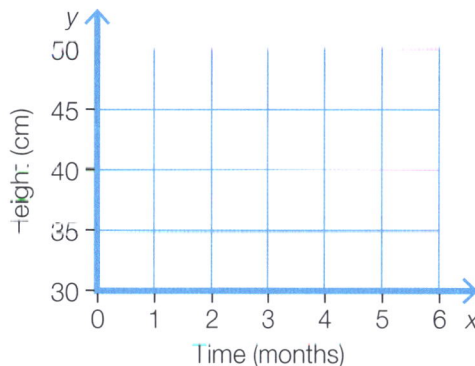

Brain-teaser The chart shows a baby's height for the first 6 months. Month '0' is the day the baby was born.

Draw a line graph to show how the baby's height has changed.

137

English glossary

A

adjectives are sometimes called 'describing words' because they pick out features of nouns such as size or colour. They can be used before or after a noun. The red bus.

adverbs can describe the manner, time, place or cause of something. They tell you more information about the event or action.

adverbials are words or phrases that give us more information about an event or action. They tell you how, when, where or why something happened.

apostrophes:
- show the place of missing letters (**contraction**)
- show who or what something belongs to (**possession**).

C

clauses are groups of words that must contain a subject and a verb. Clauses can sometimes be complete sentences.
- A **main clause** contains a subject and verb and makes sense on its own.
- A **subordinate clause** needs the rest of the sentence to make sense. A subordinate clause includes a conjunction to link it to the main clause.

commas have different uses, including:
- to separate items in a list
- to separate a fronted adverbial from the rest of the sentence
- to clarify meaning.

common nouns name something in general (boy, man).

conjunctions link two words, phrases or clauses together. There are two main types of conjunction.
- **co-ordinating conjunctions** (and, but) link two equal clauses together.
- **subordinating conjunctions** (when, because) link a subordinate clause to a main clause.

consonants are most of the letters of the alphabet except the vowel letters a, e, i, o, u.

contraction a shortened word with an apostrophe to show show the place of missing letters

co-ordinating conjunctions (and, but) link two equal clauses together.

D

determiners go before a noun (or noun phrase) and show which noun you are talking about.

direct speech is what is actually spoken by someone. The actual words spoken will be enclosed in **inverted commas**: "Please can I have a drink?"

E

exclamation marks show the end of exclamations and some commands.

F

fronted adverbials are at the start of a sentence. They are usually followed by a comma.

full stops mark the end of statements.

H

homophones are words that sound the same but are spelled differently and mean different things.

I

inverted commas are punctuation used with direct speech: "Please can I have a drink?"

M

main clause contains a subject and verb and makes sense on its own.

N

nouns are sometimes called 'naming words' because they name people, places and things. A **proper** noun (Ivan, Wednesday) names something specifically and starts with a capital letter. A **common** noun (boy, man) names something in general.

noun phrases are phrases with nouns as their main word and may contain adjectives or prepositions. Enormous grey elephant/in the garden.

P

past tense verbs describe past events. Most verbs take the suffix ed to form their past tense.

perfect form of a verb usually talks about a past event and uses the verb have + another verb. **Past perfect**: He had gone to lunch. **Present perfect**: He has gone to lunch.

phrases are groups of words that are grammatically connected so that they stay together, and that expand a single word. Phrases do not contain a subject or a verb.

plural means 'more than one'.

possession a word using an apostrophe to show who or what something belongs to.

prefix is a set of letters added to the beginning of a word in order to turn it into another word.

prepositions link nouns (or pronouns or noun phrases) to other words in the sentence. Prepositions usually tell you about place, direction or time.

present tense verbs describe actions that are happening now.

progressive or 'continuous' form of a verb describes events in progress.
- **present progressive:** We are singing.
- **past progressive:** We were singing.

pronouns are short words used to replace nouns (or noun phrases) so that the noun does not need to be repeated.
- **personal pronouns** replace people or things.
- **possessive pronouns** are used to show who something belongs to.

proper nouns name something specifically and starts with a capital letter (Ivan, Wednesday).

Q

question marks show the end of questions.

R

root word is a word to which new words can be made by adding prefixes and suffixes: happy – unhappy – happiness.

S

singular means 'only one'.

subordinate clause needs the rest of the sentence to make sense. A subordinate clause includes a conjunction to link it to the main clause.

subordinating conjunctions (when, because) link a subordinate clause to a main clause.

suffix is a word ending or a set of letters added to the end of a word to turn it into another word.

syllable sounds like a beat in a word. Longer words have more than one syllable.

T

tense is **present** or **past** tense and normally shows differences of time.

V

verbs are doing or being words. They describe what is happening in a sentence. Verbs come in different tenses.

vowel sounds are made with the letters a, e, i, o, u. Y can also represent a vowel sound.

W

word families are normally related to each other by a combination of letter pattern, grammar and meaning: child – children – childish – childlike.

Maths glossary

12-hour clock Uses 12 hours, with am before 12 noon, and pm after.

24-hour clock Uses 24 hours for the time; does not need am or pm, 17:30 = 5:30pm.

2D Two-dimensional, a flat shape like a square.

A

Acute angle An angle smaller than one right angle.

Adjacent Near or next to something, usually used for talking about angles, sides or faces.

Analogue clock Shows the time with hands on a clock face.

Angle A measure of turn, for example a right angle.

Anti-clockwise Rotating in the opposite direction to the hands of a clock.

Approximate A number found by rounding or estimating.

Area The amount of surface covered by a shape.

Axis (plural axes) The horizontal and vertical lines on a graph.

B

Bar chart A chart that uses bars of different heights to represent data.

C

Clockwise Rotating in the same direction as the hands of a clock.

Column addition/subtraction Arranging numbers in place-value columns to make addition easier.

Coordinates Numbers that give the position of a point on a coordinate grid, (x, y).

D

Decimal A number less than 1, shown with digits and a decimal point.

Decimal places The numbers to the right of the decimal point, such as tenths, hundredths.

Decimal point The dot used to separate the whole number part of a decimal from the part that is less than 1.

Denominator The number on the bottom of a fraction.

Difference The amount between two numbers.

Digits Our number system uses ten digits, 0–9, to represent all our numbers.

Digital clock Shows time using digits rather than by having hands on a clock face.

E

Equivalent fractions The same amount represented differently, such as $\frac{1}{2}$ and $\frac{2}{4}$.

Estimate To use information to get an approximate answer.

Even numbers Numbers that can be divided by 2. They end in 0, 2, 4, 6 or 8.

I

Irregular polygon A 2D shape which does not have identical sides and angles.

L

Line of symmetry Parts of a symmetrical shape are mirror images of each other either side of a line symmetry.

M

Mental methods Approaches for accurately solving calculations without writing them down.

N

Negative number A number less than zero.

Numerator The top number of a fraction. The numerator is divided by the denominator.

O

Obtuse angle An angle larger than one right angle and smaller than two right angles.

Odd numbers Numbers that cannot be divided by 2. They end in 1, 3, 5, 7 or 9.

P

Partitioning Breaking down numbers into 1000s, 100s, 10s and 1s, to make calculations easier

Perimeter The distance around the edge of a closed shape.

Pictogram A chart that uses pictures to represent data.

Q

Quadrilateral A flat shape with four sides, such as rhombus, square, rectangle, kite, parallelogram, trapezium.

R

Regular shape A 2D shape with all sides the same length and all angles the same size.

Right angle A quarter turn.

Roman numerals The system of letters used by the Romans to represent numbers.

Rounding Simplifying numbers to the nearest 10, 100 and so on.

S

Symbol A sign used for an operation or relationship in mathematics, such as +, −, ×, ÷, =, < or >.

Symmetrical A symmetrical shape is one that is identical either side of a line of symmetry.

T

Time graph Used to show how something changes over time, such as height, temperature or speed.

Translation Moving the coordinates of points or shapes by the same amounts on a graph.

Triangle A 2D shape with three sides. Can be equilateral, isosceles, right-angled or scalene.

English answers

GRAMMATICAL WORDS

Page 10

1 **a.** The <u>astronauts</u> prepared for their <u>journey</u> to (Mars.)
   b. (Ms Green) gave the <u>class</u> their <u>homework.</u>
   c. The <u>doctor</u> used a <u>stethoscope</u> to listen to (Amelia's) <u>heart.</u>
   d. The <u>tourists</u> visited (Buckingham Palace) in (London.)
   e. "My <u>birthday</u> is in (June)," said (Hannah) excitedly.

Page 11

1 Any 'interesting' adjective that makes sense in the sentence. For example:
   a. huge, gigantic, massive, enormous
   b. ancient, decaying, historic, shabby
   c. delicious, tasty, yummy, mouth-watering

2 Any adjectives (one or more, before either or both nouns) used/combined in a way that makes sense in the sentence. For example:
   a. owl: wise, old, magnificent, white, powerful, beautiful
   branch: long, strong, skinny, crooked, drooping
   b. cyclist: old, young, fit, fast, slow, wobbly, colourful, famous
   lane: bumpy, narrow, wide, winding, twisty, country
   c. teacher: old, new, grumpy, friendly, horrid, lovely, strict
   hall: old, new, sports, grand, huge, small

Page 12

1 unhelpful, incomplete, irresponsible

2 Any answer that implies the prefix 'un' has the opposite (or negative) effect on the word.

Page 13

1 **a.** <u>The lonely, frightened evacuee with a suitcase</u> stood on the platform.
   b. The robin is stood on <u>the broken, empty bird-bath by the path.</u>
   c. The children played happily in <u>the soft, yellow sand near the dunes.</u>
   d. Amber borrowed <u>the only English dictionary in the library.</u>
   e. Omar took <u>the last apple muffin on the tray.</u>

Page 15

1 **a.** We **swam** at the outdoor pool in town.
   b. He **wrote** a letter of complaint.
   c. The frog **jumped** out of the pond.

2

Sentence	Verb type
I have drawn a picture.	Past tense
I am drawing a picture.	Present tense
I draw a picture.	Past progressive
I was drawing a picture.	Present progressive
I drew a picture.	Present perfect

3 **was playing   was cooking**

4 The princess **has rescued** the prince from the tower.

Page 17

1 **a.** Sadly   **b.** sometimes   **c.** downstairs
   d. yesterday   **e.** Suddenly

2 Any adverb that makes sense in the sentence. For example:
   a. Yesterday, Happily
   b. loudly, angrily
   c. Later, Tomorrow
   d. sleepily, comfortably
   e. Carefully, Gently, Quickly

3 Any suitable adverbs. For example:
   a. unexpectedly, today, yesterday, upstairs
   b. outside, outdoors, happily, excitedly, carefully
   c. Happily, Now, Regularly, Sometimes, Consequently

Page 18

1 **a.** I went to the park <u>last Thursday afternoon.</u>
   b. We waited for our drinks <u>in the sunshine.</u>
   c. The little girl ran to the finishing line <u>as fast as she could.</u>
   d. The children were playing football <u>all morning.</u>
   e. The enormous dog barked <u>in the garden.</u>

Page 19

1 **a. In assembly,** Ms Wilkinson played the piano.
   b. Suddenly, the bell rang.
   c. At the end of break time, we went inside.
   d. Along the beach, the woman walked her dog.

Page 21

1

	Main clause	Subordinate clause
I washed my hands **after I went to the toilet.**		✓
When I lost my favourite teddy, I was upset.		✓
I jumped when the door slammed loudly.	✓	
Before I went on stage, **I was feeling nervous.**	✓	
I shut the window when it rained.	✓	
The lights went out **because the power was cut off.**		✓

2 **a.** <u>The girl walked to school</u> although it was raining.
   b. <u>The bus was late</u> because it broke down.
   c. As it was snowing, <u>the football match was cancelled.</u>

3 **a.** He cleaned out the guinea pigs <u>after feeding the rabbit.</u>
   b. <u>When we were younger,</u> we went ice-skating with our grandma.
   c. <u>If I go on Saturday,</u> I will see the animals at the zoo.

Page 22

1 **a.** She put sun cream on <u>before</u> she went outside.
   b. I have two brothers (so) I know lots about football.
   c. You can have raisins (or) you can have grapes.
   d. <u>If</u> you cook dinner, I'll do the washing up.

Page 23

1 **a.** a   **b.** your

2 **a.** The boy ate **an** orange for his lunch.
   b. Please can I have **some** peas?

Page 25

1 **a.** They   **b.** I   **c.** his   **d.** She her   **e.** she her

2 **a.** she   **b.** he   **c.** its   **d.** they   **e.** he his

Page 26

1 **a.** after   **b.** during   **c.** beside   **d.** under

2 **a.** School was closed **because of** the snow.
   b. I had to eat all my carrots **before** pudding.
   c. The snake was curled **around** a branch of the tree.

PUNCTUATION

Page 27

1 **a.** **U**sually in **O**ctober, the leaves fall off the trees.
   b. **L**ily and **M**eg visited **E**dinburgh **C**astle, on their school trip.
   c. **T**he new pilot, who was called **T**om, often flew to **G**ermany.

2 No. Capital letters are also used for proper nouns (such as names of people and places).

Page 29

1 **a.** When it rains, our garden is full of puddles.
   b. Any answer that suggests: This sentence is a statement so it needs a full stop.

2 **a.** When can I go outside to play**?**
   b. When it rains hard, we have break time inside**.**
   c. Pick that up**.** (or **!**)
   d. What an amazing feeling**!**

3 It was the school holidays**.** Jessica and Nathan were out walking in the woods when they came to a broken bridge. "How annoying**!**" cried Nathan.
   "How are we going to get across the stream now**?**" thought Jessica. They both looked at the fast-flowing stream and the slippy rocks underneath the broken bridge**.** It was no good, they would have to walk downstream until they found a safe place to cross.

Page 30

1 **a.** doesn't He's   **b.** can't I'll

2

Words in full	Contraction
you had/would	you'd
are not	aren't
should have	should've
he will	he'll

Page 31

1 **a.** children's   **b.** bridesmaids'   **c.** babies'   **d.** boys'

Page 33

1 "Look at all that rain!" exclaimed Grandad. "I think we will have to go in the car today."

2 Charlie was standing at the end of the dinner queue.
   "I am so hungry!**"** he moaned.
   "Me too. Why are we always last?**"** said his friend Sam.
   "I just hope there is some chocolate cake left,**"** replied Jing, who was just in front of Charlie.
   Then the lunchtime assistant told them, **"**You don't need to worry. There's plenty of cake for everyone.**"**

Page 34

1 He used flour, sugar, butter and eggs to make a delicious cake.

2 **a.** For breakfast, I had pancakes**,** yoghurt**,** fruit and honey.
   b. On sports day, she competed in the egg and spoon race**,** the skipping race and the obstacle race.
   c. In the film about nocturnal animals there were owls**,** bats and foxes.

Page 35

1 **a.** Before school**,** I had a swimming lesson.
   b. Last year**,** my teacher was Mr Davies.
   c. At the weekend**,** her aunt came to visit.

2 **a.** **In January**, it snowed and snowed.
   b. **Yesterday afternoon**, I went to the museum.
   c. **At the airport**, we had to sit and wait for Granny's plane.
   (The comma must be present and in the correct place.)

Page 37

1 **a.** Two clear lines (**/**): one between 'its head.' and 'After that'; one between 'in class.' and 'When we'.
   b. Any answer that suggests: the topic changes each time.
   c. first paragraph: dragonflies; second paragraph: flatworms; third paragraph: what the children did with the creatures they had found

Page 38

1 **a.** Introduction, Punch and Judy, Buckets and spades, The sea
   b. They are in bold (and on a separate line).
   c. Any answer that suggests: The sea is mentioned but the paragraph is not mainly about this.

VOCABULARY

Page 39

1

electric	music	attend	regular	legal
electricity electrician	musical musician	attention attentive	irregular regulate	illegal legality

Page 40

1

Word	New word
patient	**impatient**
responsible	**irresponsible**
act	**interact**
legible	**illegible**
marine	**submarine**

Page 41

1 finally, dangerous, politician, dramatically, preparation, invention

SPELLING

Page 43

1

Singular	Plural
boy	boys
curtain	curtains
pony	ponies
ditch	ditches
fish	fish/fishes
sheep	sheep
class	classes
life	lives

2

Singular	Plural
wish	wishes
tomato	tomatoes
plate	plates
foot	feet
fox	foxes
loaf	loaves
kiss	kisses
coin	coins

3 **a.** knives   **b.** butterflies   **c.** children

Page 45

1

air ear are	While we waited for my mum to have her h(air)cut, I sh(are)d a p(ear) with my brother. We read a book about a big brown b(ear). We took it in turns to turn the pages so that it was f(air).
ore or au aw	This m(or)ning I was dr(aw)ing a dinos(au)r when, after a sh(or)t time, my dad asked if I wanted m(or)e breakfast.
ei eight ey	My n(eigh)bours who live at number (eigh)t have a dog. He is taken for walks around the park. But the dog doesn't ob(ey) his owners so th(ey) keep him on a tight r(ei)n.
ough ow	Alth(ough) it had sn(ow)ed heavily, school was still open. Later, the caretaker sh(ow)ed us where the melted ice was fl(ow)ing down the hill.

2 **a.** autumn   **b.** pair   **c.** sore   **d.** eight   **e.** although

Page 47

1 **'k' sound:** came, caterpillar, carry, continue, candle
   's' sound: centre, city, cylinder, cycle, centipede

2 **'ch' sound:** chair, church, chocolate, cheese
   'k' sound: chemist, character, chorus, chaos
   'sh' sound: machine, brochure, chef, chalet

Page 48

1 **a.** I went on an exciting adven**ture**.
   b. The tea**cher** handed out a sticker at the end of the day.
   c. I painted a beautiful pic**ture**.
   d. The pirate was searching for trea**sure**.
   e. We went on a na**ture** trail to find bugs and worms.

Page 49

1 **a.** main   **b.** whether   **c.** heir
   d. grate   **e.** alter   **f.** fair

Page 51

1

Word	Number of syllables
ordinary	4
famous	2
natural	3
perhaps	2
interest	3
disappear	3

2 Check the children's answers.

READING

Page 53

1 **a.**
   A huge amount of waste material is produced...
   ...waste is taken to landfill sites...
   ...harmful for the environment.
   Recycling means waste products are turned into something new...
   Reusing means using items again...
   Reducing waste involves using fewer materials in the first place...
   b. Main points from the text, brief and concise, covering all 4 paragraphs. For example:
   • Huge amounts of waste in UK
   • Waste taken to landfill sites/harmful to environment
   • Recycling turns waste into something new
   • Reusing uses things again
   • Reducing uses less to start with

Page 55

1 **a.** Myth   **b.** The Minotaur   **c.** Theseus
   d. To kill the Minotaur/survive the maze
   e. Journeys or quests

Page 57

1 **a.** Three
   b. Any two: paving stones, kitchen worktops, gravestones
   c. Chalk
   d. Metamorphic rocks are created when one type of rock changes to another due to heat and pressure.

Page 59

1 **a.** It was stolen.
   b. An answer including any of the following clues:
       The boys outside the shop looked suspicious.
       The boys were whispering about the bike.
       Ollie didn't lock the bike.
       The boys had gone and so had the bike.

Page 61

1 No. Answers must refer to evidence in the text. For example, the text says:
   • she ate everything *except* the peas, which suggests she didn't like them
   • she lifted one pea slowly to her mouth/she made a face when she went to eat the pea, which suggests she didn't want to eat it.

2 **a.** Excited/happy (or similar)
   b. Answers must refer to evidence in the text. For example, the text says:
   • he ran to the door and searched the post/ripped open the tickets, which suggests he couldn't wait
   • he smiled when he saw the tickets, which suggests he was very pleased/happy.

Page 63

1 **a.** buzzed
   b. Any answer that suggests:
   It describes/gives a picture of the sound.
   c. bee buzzed/fabulously fresh fruit
   d. peacefully, slowly, really, already, noisily, fabulously
   e. Any answer that suggests:
   The sunlight filled the room (like water flooding a container).

Page 65

1 Somewhere to live

2 moved

3 **a.** Journey
   b. Answers must refer to evidence in the text. For example, the text says they went on a coach and travelled for a long time to the seaside.

Page 67

1 **a.** (1) heading, (2) subheading, (3) bullet points, (4) picture, (5) caption
   b. To emphasise the word/make the word stand out
   c. Any answer that suggests:
   The numbers show the order in which to carry out the steps.

Page 69

1 **a.** Any answer that suggests:
   It creates a shape similar to what the poem is about – a snake./The shape of the poem (a long line) emphasises the meaning and words in the text.
   b. Any answer that suggests one or both of the following:
   • The word 'snake' is bigger and in bold.
   • The red '<' looks like a forked tongue.

2 **a.** Bold writing, italics
   b. Any answers that suggest the following: it helps the reader/the actors know what the characters are doing as well as what they are saying.

Maths answers

NUMBER AND PLACE VALUE

Page 73

1. **a.** seven thousand, three hundred and eighty
   b. two thousand and sixty-nine
2. **a.** 6841 **b.** 5002
3. 8 92 250 725 1612 3875 5000 9999
4.

1000 more	3350	2243	5789	8000	9999
Number	2350	1243	4789	7000	8999
1000 less	1350	243	3789	6000	7999

Brain-teaser: **a.** Dipton **b.** Blinkton
Brain-buster: eleven thousand, two hundred

Page 75

1.

	Nearest 10	Nearest 100	Nearest 1000
a. 77	80	100	0
b. 583	580	600	1000
c. 1232	1230	1200	1000
d. 3765	3770	3800	4000

2. **a.** 80 **b.** 70 **c.** 120
3. **a.** 600 **b.** 300 **c.** 1300
4. **a.** 6000 **b.** 4000 **c.** 12,000

Brain-teaser: Blinkton and Mumsford
Brain-buster: 11,000

Page 77

1. **a.** 42, 48, 54, 60 **b.** 63, 70, 77, 84 **c.** 54, 63, 72, 81
   d. 375, 400, 425, 450 **e.** 3000, 4000, 5000, 6000
2. **a.** 78, 72, 66, 60 **b.** 63, 56, 49, 42 **c.** 63, 54, 45, 36
   d. 850, 825, 800, 775 **e.** 8000, 7000, 6000, 5000

Brain-teaser: Joe has £3 more.
Brain-buster: 32 weeks

Page 79

1. **a.** 1 **b.** –1 **c.** –3 **d.** –5 **e.** –5 **f.** –5 **g.** –1 **h.** –6
2. **a.** 2 **b.** 6 **c.** 6

Brain-teaser: –1°C
Brain-buster: 15°C

Page 81

1. **a.** IV **b.** XI **c.** XXV **d.** XIX **e.** LII **f.** XLV **g.** XC **h.** LXXXVII
2. **a.** 6 **b.** 9 **c.** 17 **d.** 22 **e.** 55 **f.** 40 **g.** 88 **h.** 90
3. 0

Brain-teaser: 14 + 23 = 37
Brain-buster: 91 – 65 = 26

CALCULATIONS

Page 83

1. **a.** 50 **b.** 89 **c.** 108 **d.** 394 **e.** 23 **f.** 86 **g.** 95 **h.** 564
2. **a.** 14 **b.** 42 **c.** 44 **d.** 151 **e.** 36 **f.** 61 **g.** 46 **h.** 238
3. **a.** 77 **b.** 99 **c.** 485 **d.** 5838 **e.** 62 **f.** 21 **g.** 133 **h.** 2323

Brain-teaser: **a.** £99 **b.** 105 miles
Brain-buster: £2300

Page 85

1. **a.** 718 **b.** 1003 **c.** 3911 **d.** 7584
2. **a.** 602 **b.** 726 **c.** 8016 **d.** 9619

Brain-teaser: 1850
Brain-buster: 17,014

Page 87

1. **a.** 218 **b.** 425 **c.** 162 **d.** 2818
2. **a.** 179 **b.** 449 **c.** 2605 **d.** 2487

Brain-teaser: **a.** 736 **b.** 111
Brain-buster: 4104

Page 89

1. **a.** 21 **b.** 45 **c.** 32
2. **a.** 4 **b.** 12 **c.** 6
3. 0 6 12 18 24 30 36 42 48 54 60 66 72
4. 0 12 24 36 48 60 72 84 96 108 120 132 144
5. 0 12 24 36 48 60 72

Brain-teaser: 42
Brain-buster: You can say 6 × 9 = 54, and then multiply by 10.
Answer = £5.40

Page 91

1. **a.** 72 **b.** 24 **c.** 950 **d.** 300 **e.** 720 **f.** 1200 **g.** 2400
   h. 9000
2. **a.** 13 **b.** 8 **c.** 15 **d.** 30 **e.** 12 **f.** 3 **g.** 20 **h.** 8

Brain-teaser: **a.** 2400p or £24 **b.** 12
Brain-buster: **a.** 7200p or £72 **b.** 4

Page 93

1. **a.** 114 **b.** 914 **c.** 1368 **d.** 4030
2. **a.** 459 **b.** 534 **c.** 2152 **d.** 7638

Brain-teaser: £10.01 or 1001p
Brain-buster: £33.75 or 3375p

Page 95

1. **a.** 25 **b.** 24 **c.** 18 **d.** 133 **e.** 1167 **f.** 107
2. **a.** 58 **b.** 43 **c.** 142 **d.** 154 **e.** 62 **f.** 103

Brain-teaser: 157
Brain-buster: 552 metres each hour

FRACTIONS AND DECIMALS

Page 97

1. **a.** $\frac{2}{6}$ shaded **b.** $\frac{3}{4}$ shaded **c.** $\frac{4}{10}$ shaded **d.** $\frac{9}{12}$ shaded
2. $\frac{1}{2} = \frac{6}{12}$, $\frac{1}{3} = \frac{4}{12}$, $\frac{1}{4} = \frac{3}{12}$, $\frac{1}{6} = \frac{2}{12}$
3. **a.** $\frac{6}{8}$ **b.** $\frac{6}{9}$ **c.** $\frac{10}{16}$ **d.** $\frac{12}{20}$

Brain-teaser: Jane
Brain-buster: Joe

Page 99

1. **a.** $\frac{1}{2}$ **b.** $\frac{1}{4}$ **c.** $\frac{2}{3}$ **d.** $\frac{4}{7}$
2. **a.** $\frac{1}{2}$ **b.** $\frac{3}{5}$ **c.** $\frac{1}{8}$ **d.** $\frac{7}{20}$
3. **a.** $\frac{5}{4}$ **b.** $\frac{7}{5}$ **c.** $\frac{9}{10}$ **d.** $\frac{11}{8}$ **e.** $\frac{11}{7}$ **f.** $\frac{12}{6}$ or 2
4. **a.** $\frac{4}{3}$ **b.** $\frac{3}{6}$ or $\frac{1}{2}$ **c.** $\frac{6}{8}$ **d.** $\frac{6}{4}$ **e.** $\frac{3}{5}$ **f.** $\frac{13}{20}$

Brain-teaser: $\frac{1}{12}$
Brain-buster: $\frac{11}{20}$

Page 101

1. **a.** $\frac{7}{10}$ **b.** $\frac{12}{10}$ **c.** $\frac{22}{10}$ **d.** $\frac{38}{100}$ **e.** $\frac{81}{100}$ **f.** $\frac{180}{100}$
2. **a.** $\frac{5}{10}$ **b.** $\frac{2}{10}$ **c.** $\frac{14}{10}$ **d.** $\frac{10}{100}$ or $\frac{1}{10}$ **e.** $\frac{52}{100}$ **f.** $\frac{45}{100}$
3. **a.** six tenths **b.** nine tenths **c.** fourteen hundredths
   d. ninety-one hundredths
4. **a.** $\frac{7}{10}$ **b.** $\frac{13}{10}$ **c.** $\frac{35}{100}$ **d.** $\frac{2}{100}$

Brain-teaser: 3 tenths = 30 hundredths
Brain-buster: 63 hundredths = 6 tenths and 3 hundredths

Page 103

1. **a.** 0.5 **b.** 0.25 **c.** 0.75
2. **a.** 0.5 **b.** 0.75 **c.** 0.1 **d.** 0.27 **e.** 0.25 **f.** 0.8
3. **a.** $\frac{1}{4}$ **b.** $\frac{78}{100}$ **c.** $\frac{4}{10}$ **d.** $\frac{3}{4}$ **e.** $\frac{1}{2}$ **f.** $\frac{21}{100}$

Brain-teaser: They are both two tenths.
Brain-buster: She is right, they are both equivalent to 0.75.

Page 105

1. **a.** 0.7 **b.** 0.31 **c.** 0.3 **d.** 0.94
2. **a.** 5 **b.** 3 **c.** 6 **d.** 7
3. **a.** 0.7 < 0.75 **b.** 0.31 < 0.42 **c.** 0.6 = 0.60 **d.** 0.25 > 0.23
4. Decimals should be arranged in this order: 0.05, 0.25, 0.35, 0.5, 0.65, 0.75, 0.9

Brain-teaser: 6.5
Brain-buster: 0.6

MEASUREMENT

Page 107

1. **a.** centimetres **b.** metres **c.** pence **d.** pounds **e.** millilitres
   f. litres **g.** kilograms **h.** grams **i.** minutes or seconds
   j. days or weeks
2. **a.** weighing scales **b.** measuring cylinder **c.** stopwatch
3. **a.** ruler **b.** tape measure **c.** measuring cylinder
   d. weighing scales **e.** stopwatch **f.** calendar

Brain-teaser: 5000g
Brain-buster: 865mm

Page 109

1. 2 years = 730 days; 2 weeks = 14 days; 2 days = 48 hours;
   2 hours = 120 minutes; 2 minutes = 120 seconds
2. **a.** 120 seconds **b.** 180 minutes **c.** 96 hours **d.** 35 days
   e. 72 months **f.** 730 days
3. **a.** 210 seconds **b.** 150 minutes **c.** 132 hours **d.** 90 months

Brain-teaser: 3653 days
Brain-buster: **a.** 8760 **b.** 1440

Page 111

1.

Analogue	twelve noon	quarter to nine	ten past eleven	five to four	quarter past three
Digital	12.00	8.45	11.10	3.55	3.15

2. **a.** 01:50 **b.** 16:25 **c.** 06:00 **d.** 23:15
3. **a.** half past eleven am or 11.30am
   b. quarter past three pm or 3.45pm
   c. twenty-five past three am or 3.25am
   d. quarter to one pm or 12.45pm

Brain-teaser: 1 hour and 20 minutes
Brain-buster: 11 hours and 40 minutes

Page 113

1.

Pence	500p	150p	3300p	59p	1000p
Pounds	£5	£1.50	£33.00	£0.59	£10.00

2.

Pounds	£1	£4.25	£0.62	£20	£12.06
Pence	100p	425p	62p	2000p	1206p

3. **a.** £5.80 **b.** £8.10 **c.** £3.50 **d.** £7.01 **e.** £4 **f.** £10
   e. £5 **h.** £5

Brain-teaser: **a.** £6.25 **b.** £3.75
Brain-buster: **a.** £14.25 **b.** £5.75

Page 115

1. thimble = 20ml, mug = 200ml, bathtub = 200l
2. mouse = 50g, child = 50kg, elephant = 5000kg
3. **a.** 5000g **b.** 6kg **c.** 500g **d.** $4\frac{1}{2}$kg or 4.5kg
4. **a.** 3l **b.** 7500ml **c.** $3\frac{1}{2}$l or 3.5l **d.** 500ml
5. **a.** $3\frac{3}{4}$kg or 3.75kg **b.** 1150g **c.** 420g **d.** 3700ml **e.** 0.37l
   f. 5000ml

Brain-teaser: **a.** 20 **b.** 2kg and 700g
Brain-buster: **a.** 25 **b.** 8

Page 117

1. **a.** 50mm **b.** 30mm **c.** 45mm **d.** 27mm
2. **a.** 4cm **b.** 7cm **c.** 6.3cm
3.

a.

mm	cm
10	1
100	10
20	2
350	35
1000	100

b.

cm	m
100	1
1000	10
25	0.25
50	0.5
1000	10

c.

m	km
500	$\frac{1}{2}$
2000	2
250	$\frac{1}{4}$
1000	1
9000	9

Brain-teaser: 1.54m or 154cm
Brain-buster: $9\frac{1}{2}$km or 9.5km or 9500m

Page 119

1. **a.** length 4cm, width 2cm
      perimeter = 12cm
   b. side length 3cm
      perimeter = 12cm
2. **a.** 8cm **b.** 8cm **c.** 12cm
3.

Shape	Length	Height	Perimeter
rectangle	5cm	2cm	**14cm**
rectangle	12mm	5mm	**34mm**
rectangle	6km	2km	**16km**
square	8mm	8mm	**32mm**
square	5m	5m	**20m**
square	4.5cm	4.5cm	**18cm**

Brain-teaser: 14km

Page 121

1 **a.** 6 squares   **b.** 16 squares   **c.** 5 squares   **d.** 12 squares

2 **a.** 12 squares   **b.** 9 squares

Brain-teaser: The area of the rectangle is 20 squares greater than the square.
Brain-buster: 10 blocks

GEOMETRY

Page 122

1 **a.** acute   **b.** obtuse   **c.** right angle

2 **a.** 3   **b.** 4   **c.** 2   **d.** 1

Brain-teaser: They make a straight line, which is two right angles.

Page 123

1 **a.** Right-angled   **b.** Isosceles   **c.** Equilateral   **d.** Scalene

Brain-teaser:

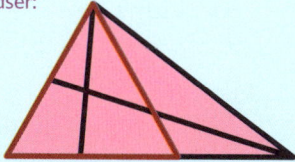

There are various possibilities. The isosceles triangle must have two sides of the same length; and the scalene must have sides that are all different lengths.

Page 125

1 A square has sides that are all the same length, whereas a rectangle only has opposite sides of equal length.

2 All the sides of a rhombus are the same length, whereas only adjacent sides are equal on a kite.

3 A parallelogram has two pairs of parallel sides, whereas a trapezium only has one pair of parallel sides.

4

a. Trapezium	**b.** Parallelogram	**c.** Kite	**d.** Rhombus

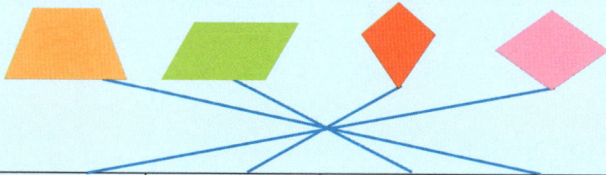

Opposite sides equal. and parallel Opposite angles equal.	Adjacent sides equal.	All sides equal. Opposite angles equal.	Only one pair of parallel sides.

Brain-teaser: A square or a rhombus
Brain-buster: It is possible. Check that the sketch shows the right angle at the top of the kite, with equal sides in the correct places.

Page 127

1

2

3

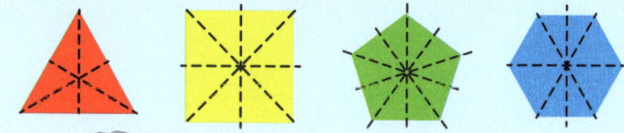

Brain-teaser:

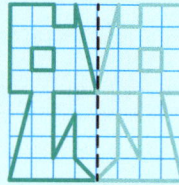

Brain-buster: H I O X

Page 129

1 **a.** X = (3, 4), Y = (7, 5)
   b. Check that a square has been plotted with these coordinates: A (2, 2), B (2, 8), C (8, 8), D (8, 2)

2 **a.** P = (2, 2), Q = (3, 6), R = (6, 3)
   b. Check that a rectangle has been plotted with these coordinates: J (1, 1), K (9, 1), L (9, 7), M (1, 7)

Brain-teaser: (7, 5)

Page 131

1 **a.** Check that the points have been plotted correctly.
   b. W = (9, 1), X = (6, 2), Y = (5, 0), Z = (4, 4)

2 **a.** Check that a triangle has been plotted correctly at these coordinates: A (2, 2), B (4, 4), C (4, 2)
   b. Check that a triangle has been plotted correctly at these coordinates: A = (5, 7), B = (7, 9), C = (7, 7)
   c. A = (5, 7), B = (7, 9), C = (7, 7)

Brain-teaser: 5 left, 4 down

STATISTICS

Page 133

1 Check that the pictogram has been drawn correctly.

2 **a.** elephant   **b.** lion and zebra   **c.** giraffe and zebra
   d. 180cm   **e.** 300kg   **f.** 6 years

Brain-teaser: lion and zebra

Page 135

1 Mariam

2 27

3 117

4 14

Brain-teaser: Check that the bar chart has been drawn correctly.

Page 137

1 4am

2 10pm

3 8°C

Brain-teaser: Check that the time graph has been drawn correctly.

Notes

English progress tracker

Grammatical words

Practised Achieved

☐	☐	Proper nouns and common nouns	10
☐	☐	Adjectives	11
☐	☐	Adjectives with prefixes	12
☐	☐	Noun phrases	13
☐	☐	Verb tenses	14
☐	☐	Adverbs	16
☐	☐	Adverbials	18
☐	☐	Fronted adverbials	19
☐	☐	Clauses	20
☐	☐	Conjunctions	22
☐	☐	Determiners	23
☐	☐	Pronouns	24
☐	☐	Prepositions	26

Punctuation

Practised Achieved

☐	☐	Capital letters	27
☐	☐	Full stops, question marks and exclamation marks	28
☐	☐	Apostrophes for contraction	30
☐	☐	Apostrophes for possession	31
☐	☐	Inverted commas	32
☐	☐	Commas in lists	34
☐	☐	Commas after fronted adverbials	35
☐	☐	Paragraphs	36
☐	☐	Headings	38

Vocabulary

Practised Achieved

☐	☐	Word families	39
☐	☐	Prefixes	40
☐	☐	Suffixes	41

Spelling

Practised Achieved

☐	☐	Plurals	42
☐	☐	Longer vowel sounds	44
☐	☐	Tricky sounds	46
☐	☐	Tricky endings	48
☐	☐	Homophones	49
☐	☐	Syllables and longer words	50

Reading

Practised Achieved

☐	☐	Identifying and summarising main ideas	52
☐	☐	Identifying themes and conventions	54
☐	☐	Retrieving and recording information	56
☐	☐	Making predictions	58
☐	☐	Making inferences	60
☐	☐	Language features	62
☐	☐	Words in context	64
☐	☐	Presentational features: non-fiction	66
☐	☐	Presentational features: fiction	68

Maths progress tracker

Number and place value

Practised	Achieved		
☐	☐	Numbers to 9999	72
☐	☐	Estimating and rounding	74
☐	☐	Counting in steps	76
☐	☐	Negative numbers	78
☐	☐	Roman numerals	80

Calculations

Practised	Achieved		
☐	☐	Mental methods for addition and subtraction	82
☐	☐	Written methods for addition	84
☐	☐	Written methods for subtraction	86
☐	☐	Times tables facts	88
☐	☐	Mental methods for multiplication and division	90
☐	☐	Written methods for short multiplication	92
☐	☐	Written methods for short division	94

Fractions and decimals

Practised	Achieved		
☐	☐	Equivalent fractions	96
☐	☐	Adding and subtracting fractions	98
☐	☐	Tenths and hundredths	100
☐	☐	Fraction and decimal equivalents	102
☐	☐	Working with decimals	104

Measurement

Practised	Achieved		
☐	☐	Units of measurement	106
☐	☐	Units of time	108
☐	☐	Analogue and digital clocks	110
☐	☐	Money	112
☐	☐	Mass and capacity	114
☐	☐	Length and distance	116
☐	☐	Perimeter	118
☐	☐	Area	120

Geometry

Practised	Achieved		
☐	☐	Angles	122
☐	☐	Triangles	123
☐	☐	Quadrilaterals	124
☐	☐	Symmetry of 2D shapes	126
☐	☐	Coordinates	128
☐	☐	Translation	130

Statistics

Practised	Achieved		
☐	☐	Tables and pictograms	132
☐	☐	Bar charts	134
☐	☐	Time graphs	136

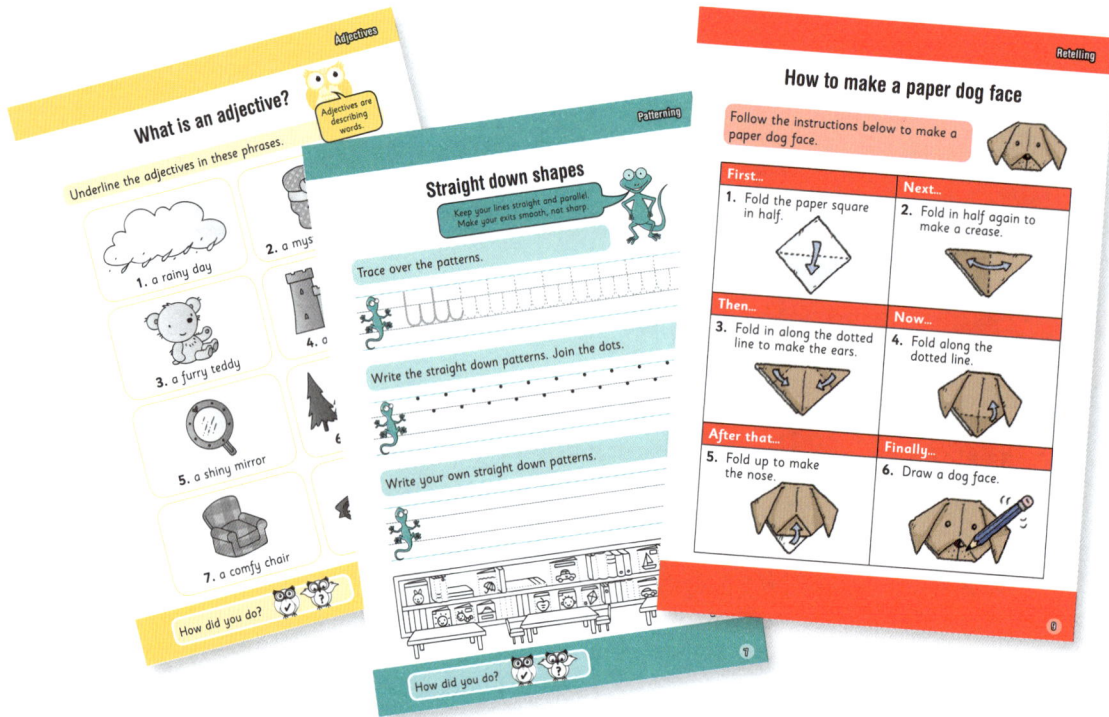